# Physiological Disorders & Mineral Deficiency of Vegetables

# Physiological Disorders & Mineral Deficiency of Vegetables

Editor

**Rachna Desai**

scitus academics

**Physiological Disorders & Mineral Deficiency of Vegetables**

Edited by **Rachna Desai**

Printed in 2017

ISBN: 978-1-68117-083-1

Library of Congress Control Number: 2015935481


SCITUS Academics LLC,
616, Corporate Way, Suite 2, 4766,
Valley Cottage, NY 10989

www.scitusacademics.com



**Notice**

# Contents

Preface ....................................................................................vii

Chapter 1 Diet, Genetics, and Disease: A Focus on the Middle East and North Africa Region ....................................................................................1
Akl C. Fahed, Abdul-Karim M. El-Hage-Sleiman, Theresa I. Farhat, and Georges M. Nemer

Chapter 2 Role of Nutrition in the Management of Hepatic Encephalopathy in End-Stage Liver Failure ....................................................................................49
Chantal Bémeur, Paul Desjardins, and Roger F Butterworth

Chapter 3 Transgenic Vegetable Breeding for Nutritional Quality and Health Benefits ....................................................................................83
João Silva Dias and Rodomiro Ortiz

Chapter 4 Magnesium and Vascular Changes in Hypertension ....................111
Ana Rosa Cunha, Bianca Umbelino, Margarida L. Correia, and Mario Fritsch Neves

Chapter 5 Structural and Functional Alterations in Photosynthetic Apparatus of Plants under Cadmium Stress ....................................................129
Pooja Parmar, Nilima Kumari, and Vinay Sharma

Chapter 6 Integrative Medicine and Human Health - the Role of Pre-, Pro- and Synbiotics Integrative Medicine and Human Health - The Role of Pre-, Pro- and Synbiotics ....................................................147
Stig Bengmark

Chapter 7 A Review on Integrated Agro-Technology of Vegetables ................183
T. S. Workneh and G. Osthoff

Chapter 8 The Importance of Mineral Elements for Humans, Domestic Animals and Plants: A Review ........................................................ 239
K. O. Soetan, C. O. Olaiya , and O. E. Oyewole

Citations ........................................................................................ 299

Index ............................................................................................. 303

# Preface

Visual nutrient deficiency symptoms can be a very powerful diagnostic tool for evaluating the nutrient status of plants. One should keep in mind, however, that a given individual visual symptom is seldom sufficient to make a definitive diagnosis of a plant's nutrient status. Many of the classic deficiency symptoms such as tip burn, chlorosis and necrosis are characteristically associated with more than one mineral deficiency and also with other stresses that by themselves are not diagnostic for any specific nutrient stress. However, their detection is extremely useful in making an evaluation of nutrient status. In addition to the actual observations of morphological and spectral symptoms, knowing the location and timing of these symptoms is a critical aspect of any nutrient status evaluation. Plants do not grow in isolation, they are part of the overall environment and as such they respond to environmental changes as that affect nutrient availability. Also, plants do influence their environment and can contribute to environmental changes, which in turn can affect the nutrient status of the plant. Stresses such as salinity, pathogens, and air pollution induce their own characteristic set of visual symptoms. Often, these symptoms closely resemble those of nutrient deficiency. Pathogens often produce an interveinal chlorosis, and air pollution and salinity stress can cause tip burn. Although at first these symptoms might seem similar in their general appearance to nutrient deficiency symptoms, they do differ in detail and/or in their overall developmental pattern. Pathological symptoms can often be separated from nutritional symptoms by their distribution in a population of affected plants. If the plants are under a nutrient stress, all plants of a given type and age in the same environment tend to develop similar symptoms at the same time. However if the stress is the result of pathology, the development of symptoms will have a tendency to vary between plants until a relatively advanced stage of the pathology is reached.

**Editor**

Chapter 1

# Diet, Genetics, and Disease: A Focus on the Middle East and North Africa Region

Akl C. Fahed[1], Abdul-Karim M. El-Hage-Sleiman[2], Theresa I. Farhat[2], and Georges M. Nemer[2]

[1]Department of Genetics, Harvard Medical School, 77 Avenue Louis Pasteur, Boston, MA 02115, USA

[2]Department of Biochemistry and Molecular Genetics, American University of Beirut, Beirut 1107 2020, Lebanon

## ABSTRACT

The Middle East and North Africa (MENA) region suffers a drastic change from a traditional diet to an industrialized diet. This has led to an unparalleled increase in the prevalence of chronic diseases. This review discusses the role of nutritional genomics, or the dietary signature, in these dietary and disease changes in the MENA. The diet-genetics-disease relation is discussed in detail. Selected disease categories in the MENA are discussed starting with a review of their epidemiology in the different MENA countries, followed by an examination of the known genetic factors that have been reported in the disease discussed, whether inside or outside the MENA. Several diet-genetics-disease relationships in the MENA may be contributing to the increased prevalence of civilization disorders of metabolism and micronutrient deficiencies. Future research in the field of nutritional genomics in the MENA is needed to better define these relationships.

# INTRODUCTION

Over the past few decades, the MENA has been witnessing significant changes in food habits paralleled by an important preponderance of metabolite-related diseases. In a region whose traditional diet is known to be healthy due to high vegetable proteins, fibers, minerals, and vitamins with low content of unfavorable food products, the "industrialization/westernization of the diet" is a well-studied and documented phenomenon [1–3]. The MENA has been losing its traditional diet which was distinguished by its diversity and richness in raw foods, proteins, and multivitamins, in the favor of a more industrial diet which consists of increased preprocessed foods, sugars, fats, alcohol, animal products, saturated- and trans-fatty acids, and relatively less vitamins and minerals with decreased consumption of milk, fruits, and vegetables [4]. A big part of this change is attributed to the lifestyle changes and globalization with the invasion of western fast food to the MENA countries. Dietary choices, minimum physical activity, religious habits, consumer ignorance, high population growth rates, economic factors, and lack of both protection laws and food fortification programs are other critical factors that influence the nutritional status in the region [5]. These changes in dietary and lifestyle patterns contribute to an increase in the rates of micronutrients deficiencies, diet-related chronic diseases, and obesity in all groups of the population in the region [5]. Due to this grave impact on chronic diseases, diet became a target of public health initiatives that aim at restoring the traditional diet of MENA countries in order to improve health conditions in their populations [6–8]. The epidemiology of diet-related diseases in the MENA region and background information on the diet-genetics-disease interaction, followed by nutrigenomic examples on diet-related diseases in the MENA region, will be discussed for the first time in this paper.

## EPIDEMIOLOGY OF DIET-RELATED DISEASES IN THE MENA

We review the numbers and trends for selected chronic metabolic diseases and micronutrient deficiencies in the different countries of the MENA where data are available (Figure 1). Given the lack of nationwide data, prevalence rates are commonly reported as estimations [9, 10]. Since 1982, the need for "direct evidence of a secular [increase]" in diseases has been established, prospecting an association with "acculturation" of traditional or rural populations [11]. Still, the numbers show alarming trends for cardiovascular diseases and metabolic disorders, namely, insulin resistance, adiposity, dyslipidemias, and atherosclerosis. Likewise, micronutrient deficiencies (MNDs) have been heavily studied recently due to their crucial contribution to the global burden of many chronic diseases.

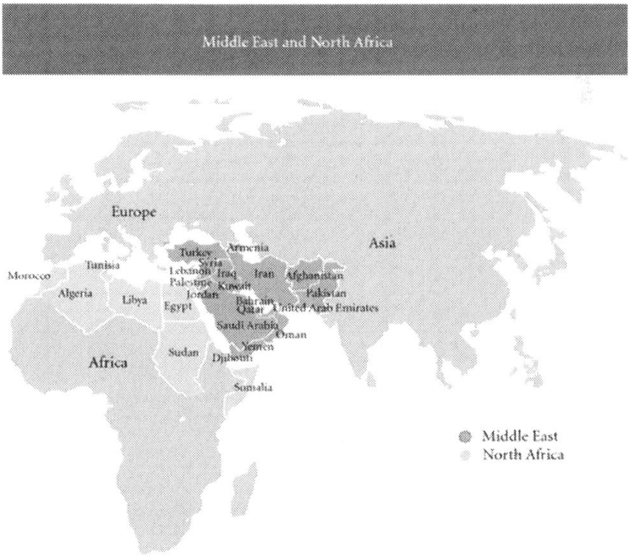


**Figure 1:** Map of the Middle East and North Africa (MENA) region. The MENA region includes countries such as Algeria, Armenia, and Turkey, that are not

members of the WHO Eastern Mediterranean Region (EMR) that is referred to in the literature.

The region is witnessing an "explosion" of Type 2 DIABETES MELLITUS (T2DM) according to the International Diabetes Foundation (IDF) (Figure 2(a)) with five countries of the MENA ranking among the top ten diabetic worldwide [15]. Similarly, the IDF reports an increase in the incidence of Type 1 DM in children of the MENA over the past decade (Figure 2(b)). The World Health Statistics of the WHO show a sharp rise in the prevalence of obesity (Figure 3), which is the most powerful and easily reported risk marker and hence the sole epidemiological WHO indicator for cardiovascular and metabolic comorbidities. In one epidemiological study, cardiovascular mortality in the MENA has been estimated to triple from 1990 to 2020 (Figure 4). Table 1 summarizes rates of cardiovascular diseases and two major comorbid risk factors, hypertension and the metabolic syndrome, with wide variations across countries of the MENA.

**Table 1:** Rates of cardiovascular disease, hypertension, and the metabolic syndrome in MENA countries from different studies

| | | | |
|---|---|---|---|
| Coronary artery disease (CAD) | | | |
| Iran | Age-adjusted prevalence (%) | 12.7 | Nabipour et al. [73] |
| Jordan | Prevalence (%) | 5.9 | Nsour et al. [74] |
| Saudi Arabia, | | | |
| (rural) | Prevalence (%) | 4.0 | Al-Nozha et al. [75] |
| (urban) | Prevalence (%) | 6.2 | |
| (overall) | Prevalence (%) | 5.5 | |
| Tunisia | Prevalence (%), [men] | 12.5 | Ben Romdhane et al. [76] |
| | Prevalence (%), [women] | 20.6 | |
| Cerebrovascular Accidents | | | |
| Bahrain | Age-adjusted incidence (per 100,000) | 96.2 | Al-Jishi and Mohan [77] |

| | | | |
|---|---|---|---|
| Iran | Age-adjusted incidence (per 100,000) | 61.5 | Ahangar et al. [78] |
| Kuwait | Age-adjusted incidence (per 100,000) | 92.2 | Abdul-Ghaffar et al. [79] |
| Libya | Age-adjusted incidence (per 100,000) | 114.2 | Radhakrishnan et al. [80] |
| Palestine | Age-adjusted incidence (per 100,000) | 62.7 | Sweileh et al. [81] |
| Qatar | Age-adjusted incidence (per 100,000) | 123.7 | Hamad et al. [82] |
| Saudi Arabia | Age-adjusted incidence (per 100,000) | 38.5 | Al-Rajeh et al. [83] |
| Hypertension (HTN) | | | |
| Algeria | Prevalence (%), [Age > 25] | 36.2 | Yahia-Berrouiguet et al. [84] |
| Bahrain | Prevalence (%), [Age > 20] | 42.1 | Al-Zurba [85] |
| Egypt | Age-adjusted prevalence (%) | 27.4 | Ibrahim et al. [86] |
| | Prevalence (%), [Age > 25] | 26.3 | Galal [87] |
| Iran | Prevalence (%), [Age > 19] | 25.6 | Sarraf-Zadegan et al. [88] |
| | Prevalence (%), [Age: 30–55] | 23.0 | Haghdoost et al. [89] |
| | Prevalence (%), [Age > 55] | 49.5 | |
| Iraq | Prevalence (%), [Age > 20] | 19.3 | WHO: STEPwise, [90] |
| Jordan | Prevalence (%), [Age > 18] | 30.2 | Zindah et al. [91] |
| Lebanon | Prevalence (%), [Age: 18–65] | 31.2 | Sibai et al. [92] |

| | | | |
|---|---|---|---|
| Morocco | Prevalence (%), [Age > 20] | 33.6 | Tazi et al. [93] |
| Oman | Prevalence (%), [Age > 20] | 21.5 | Hasab et al. [94] |
| Palestine (WB), | | | |
| (rural) | Prevalence (%), [Age: 30–65] | 25.4 | Abdul-Rahim et al. [95] |
| (urban) | Prevalence (%), [Age: 30–65] | 21.5 | |
| Qatar | Prevalence (%), [Age: 25–65] | 32.1 | Bener et al. [96] |
| Saudi Arabia | Prevalence (%), [Age: 30–70] | 26.1 | Al-Nozha et al. [97] |
| Sudan | Prevalence (%), [Age: 25–64] | 23.6 | WHO: STEPwise, [90] |
| Syria | Prevalence (%), [Age: 18–65] | 40.6 | Maziak et al. [98] |
| Turkey | Age-adjusted prevalence (%) | 25.7 | Sonmez et al. [99] |
| UAE | Prevalence (%), [Age > 20] | 20.8 | Baynouna et al. [100] |
| Yemen | Prevalence (%), [Age > 35] | 26.0 | Gunaid and Assabri [101] |
| Middle East | Prevalence (%), (overall) | 21.7 | Motlagh et al. [102] |
| Metabolic Syndrome | | | |
| Algeria | Prevalence (%), [Age > 20] | 17.4 | Mehio Sibai et al. [103] |
| Iran | Prevalence (%), [Age > 19] | 23.3 | Mehio Sibai et al. [103] |
| Jordan | Prevalence (%), [Age > 18] | 36.3 | Khader et al. [104] |
| Kuwait | Prevalence (%), [Age > 20] | 24.8 | Al Rashdan and Al Nesef [105] |
| Lebanon | Prevalence (%), [Age: 18–65] | 25.4 | Mehio Sibai et al. [103] |
| Morocco, (rural) | Prevalence (%), [women] | 16.3 | Rguibi and Belahsen [106] |
| Oman, | | | |

| | | | |
|---|---|---|---|
| (overall) | Prevalence (%), [Age > 20] | 21.0 | Al-Lawati et al. [107] |
| (Nizwa) | Age adjusted prevalence (%) | 8.0 | Al-Lawati et al. [107] |
| Palestine (WB) | Prevalence (%), [Age: 30–65] | 17.0 | Abdul-Rahim et al. [95] |
| Qatar | Prevalence (%), [Age > 20] | 27.7 | Musallam et al. [108] |
| | Age adjusted prevalence (%) | 26.5 | Bener et al. [109] |
| Saudi Arabia | Age adjusted prevalence (%) | 39.3 | Al-Nozha et al. [110] |
| Tunisia | Prevalence (%), [Age > 20] | 16.3 | Bouguerra et al. [111] |
| UAE | Prevalence (%) | 39.6 | Malik and Razig [112] |

Nonadjusted rates from different studies are not valid for comparison but displayed to present the burden of the morbidities. HTN is defined as BP > 140/90 or use of antihypertensive medications. Metabolic Syndrome definition is based on Adult Treatment Panel III, except for Palestine and Tunisia where, respectively, WHO criteria and hypercholesterolemia (Total Cholesterol ≥5.2 mmol/l) instead of low HDL cholesterol were used. UAE: United Arab Emirates; WB: West Bank [103, 113–116].

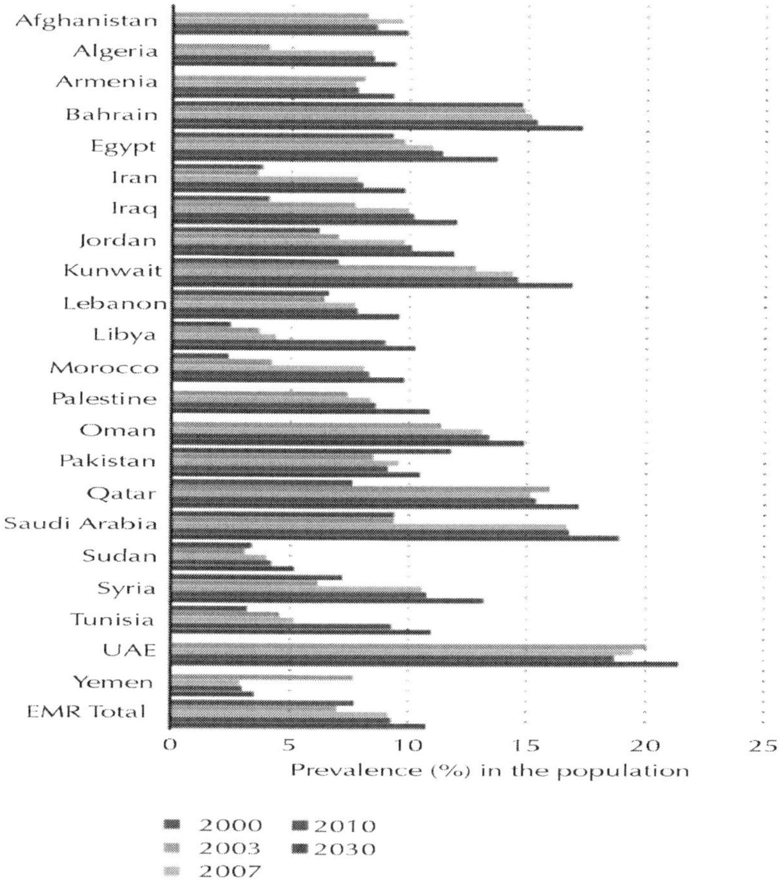


**Figure 2:** Increasing prevalence of DIABETES MELLITUS in the MENA. (a) General increase in reported prevalence (%) of Type 2 Diabetes Mellitus in the MENA between the years 2000 and 2010. Numbers are reported as approximated by the International Diabetes Federation [12–15]. The expected 2-fold increase for the year 2030 is approximated based on demographic parameters, without accounting for changes in age strata or other risk factors [16]. (b) Overall growth in annual incidence (per 100,000) of Type 1 diabetes mellitus in children younger than 14 years old in the MENA. Numbers are estimations by the International Diabetes Federation based on various years between 1986 and 2000 [12–15].

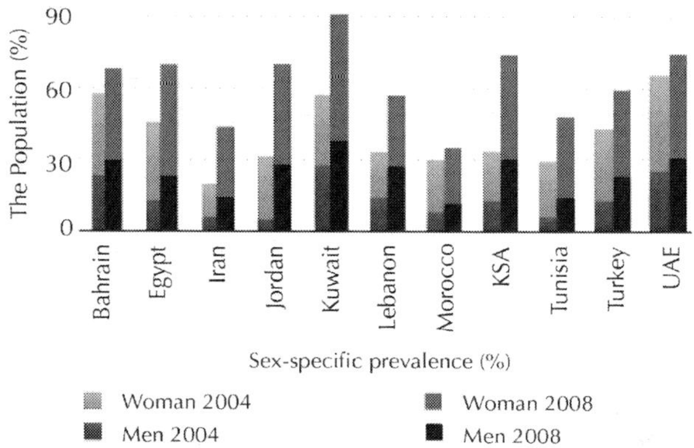


**Figure 3:** Increasing prevalence of obesity in the MENA. Prevalence (%) of obesity increased in both men and women in countries of the MENA between 2004 and 2008. Numbers are WHO estimates in World Health Statistics of 2005 and 2011. Totals of men and women are integrated for purposes of comparative illustration and do not represent adjusted arithmetic total prevalences. Obesity was defined as body mass index (BMI) ≥30 Kg/m$^2$. Obesity data about Jordanian men in 2004 are not available, but prevalence was estimated to be less than that of women [17, 18]. KSA: Kingdom of Saudi Arabia; UAE: United Arab Emirates.

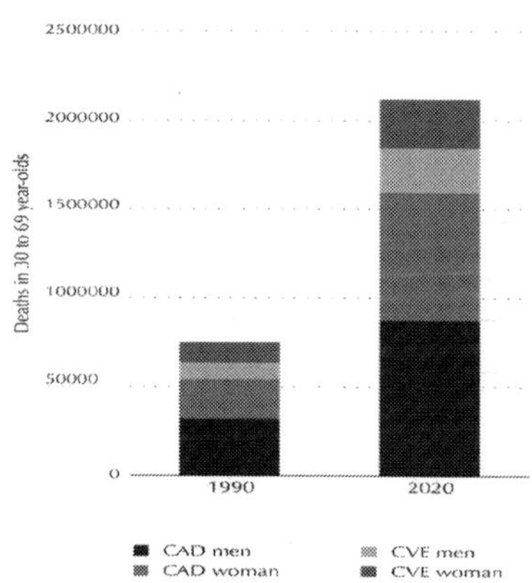


**Figure 4:** Expected overall increase in mortality due to cardiovascular diseases in the MENA [19]. CAD: Coronary Artery Disease; CVE: Cerebrovascular Event.

In the other category of diseases, deficiencies in vitamins and minerals affect particularly children and women of childbearing age. MNDs impair physical and mental development of children, exacerbate infections and chronic diseases, and impact morbidity and mortality. Most countries of the MENA have widespread MNDs, yet countries of the Gulf Cooperation Council, Iran, and Tunisia have moderate levels of MNDs [20]. Among MNDs, iron deficiency is the most prevalent nutritional problem in the MENA [21]. Its prevalence in the region varies from 17 to 70% among preschoolers, 14 to 42% among adolescents, and 11 to more than 40% among women of childbearing age. Severe iron deficiency is a direct cause of anemia. Nonnutritional genetic anemias are known to be relatively common in the region but will not be tackled because they are not affected by dietary factors. Nutritional deficiencies other than iron, such as folic acid, vitamin B12, and vitamin C, are also prevalent in some countries of the region but data are scarce [20]. The Middle Eastern and South Asian regions have the highest rates of Vitamin D Deficiency (VDD) worldwide [22]. The prevalence ranges of VDD in the MENA are 46–83% for adolescents and adults, and 50–62% among veiled women [5]. It reaches up to 70% in Iran [23] and 80% in Saudi Arabia [24]. Table 2 shows prevalence of VDD and iron deficiency in countries of the MENA where data are available.

**Table 2:** Rates of Vitamin D deficiency and iron deficiency in MENA countries from different studies

| Vitamin D deficiency (VDD) | | | |
|---|---|---|---|
| Iran | Prevalence (%), [girls], [adolescent] | Up to 70 | Moussavi et al. [143] |
| Jordan | Prevalence (%), | | |
| | [adult females] | 37.3 | Batieha et al. [144] |
| | [adult males] | 5.1 | |
| Lebanon | Prevalence (%), | | |
| | [girls] | 32 | El-Hajj Fuleihan et al. [145, 146] |
| | [boys] | 9–12 | |
| Morocco (Rabat) | Prevalence (%), [women] | 91 | Arabi et al. [147] |
| Saudi Arabia | Prevalence (%), [girls], [adolescent] | Up to 80 | Siddiqui and Kamfar [148] |

| | | | |
|---|---|---|---|
| Tunisia (Ariana) | Prevalence (%), [women], [Age: 20–60] | 47.6 | Arabi et al. [147] |
| Turkey (Ankara) | Prevalence (%), | | |
| | [mothers] | 46 | Arabi et al. [147] |
| | [newborns] | 80 | |
| Iron deficiency | | | |
| Arab Gulf countries | Prevalence (%), | | |
| | [children], [preschool age] | 20–67 | Musaiger [149] |
| | [children], [school age] | 12.6–50 | |
| | [pregnant women] | 22.7–54 | |
| Bahrain | Prevalence (%), | | |
| | [children], [Age: 6–59 months] | 48 | Bagchi [150] |
| | [women], [Age: 15–49] | 37.3 | |
| Egypt | Prevalence (%), | | |
| | [children], [Age: 6–59 months] | 25 | Bagchi [150] |
| | [women], [Age: 15–49] | 11 | |
| Iran | Prevalence (%), | | |
| | [children], [Age: 6–59 months] | 15–30 | Bagchi [150] |
| | [women], [Age: 15–49] | 33.4 | |
| Jordan | Prevalence (%),[children], [school age] | 20 | Bagchi [150] |
| | Prevalence (%), [women], [Age: 15–49] | 28 | |
| Lebanon | Prevalence (%), [children], [Age: 6–59 months] | 23 | Bagchi [150] |
| Morocco | Prevalence (%), | | |
| | [children], [Age: 6–59 months] | 35 | Bagchi [150] |
| | [women], [Age: 15–49] | 30.1 | |
| Oman | Prevalence (%), | | |
| | [children], [Age: 5–14] | 41 | Bagchi [150] |

| | | | |
|---|---|---|---|
| | [women], [Age: 15–49] | 40 | |
| Pakistan | Prevalence (%), | | |
| | [children], [Age: 6–59 months] | 60 | Bagchi [150] |
| | [women], [Age: 15–49] | 30 | |
| Palestine | Prevalence (%), | | |
| | [children], [Age: 6–59 months] | 53 | Bagchi [150] |
| | [women], [Age: 15–49] | 36.2 | |
| Saudi Arabia | Prevalence (%), [children], [preschool age] | 17 | Bagchi [150] |
| Syria | Prevalence (%), | | |
| | [children], [Age: 6–59 months] | 23 | Bagchi [150] |
| | [women], [Age: 15–49] | 40.8 | |
| UAE | Prevalence (%), | | |
| | [children], [Age: 6–59 months] | 34 | Bagchi [150] |
| | [pregnant women] | 14 | |
| Yemen | Prevalence (%), [children], [preschool age] | 70 | Bagchi [150] |

Different limits of blood levels define VDD, ranging from insufficiency to severe deficiency, similar for Iron deficiency. UAE: United Arab Emirates [147, 150].

# THE DIETARY SIGNATURE ON THE GENOME

The effect of food on gene function is the focus of nutritional genomics, an emerging field of study that focuses on the molecular, cellular, and systemic levels of this effect [25, 26]. The abundance of calories, macro- and micronutrients, and BIOACTIVE food elements constitutes the nutritional environment that alters the genome, the epigenome, the posttranscriptional regulation, and the posttranslational modifications, leading to a variety of metabolic functional gene-products, as shown in Figure 5. The nutritional environment channels the function of gene-products into certain pathways, preferring certain biological activities over others and resulting in the final phenotypic outcome and health-disease status. Correspondingly, two disciplines of nutritional genomics are entertained: nutrigenomics and nutrigenetics. The former started as a focus on the effect of nutrients on gene expression, while the latter, yet a different field of study, emerged in search of approaches to alter the clinical manifestations of certain rare diseases, such as certain inborn errors of metabolism, via personalized diet. The link between them is the functional gene product. It is the end-point in nutrigenomics and the starting point in nutrigenetics. Other conventional terminologies relevant to the diet-genetics-disease relation will also be treated in the following section, prior to discussion of the dietary signature on the genome in the MENA.

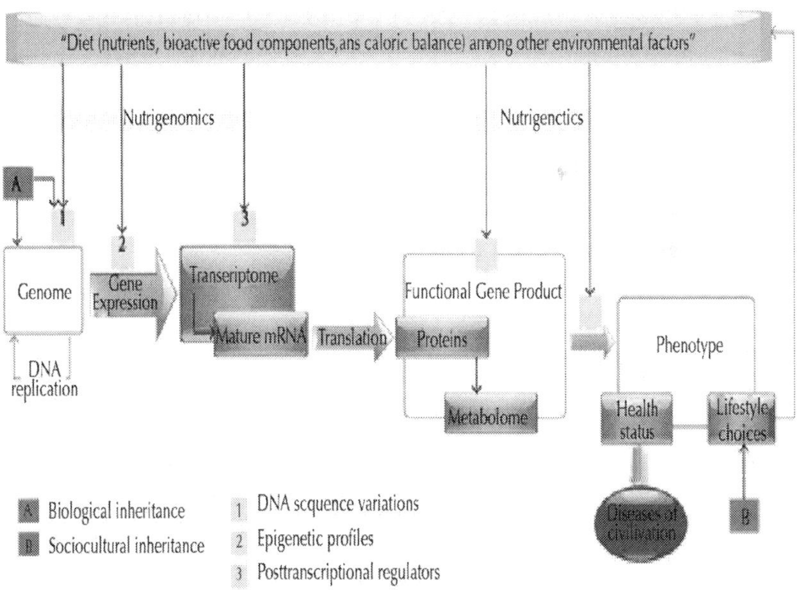


**Figure 5:** The dietary signature. The biologically inherited DNA genome accumulates DNA sequence variations over generations. Epigenetic profiles determine which parts of it are to be transcribed. Once transcribed to RNA, it matures into different mature RNA outcomes depending on the post-transcriptional regulators. Among the effects of the environment on DNA sequence

variations, epigenetic profiles, and post-transcriptional regulators, the effect of diet is studied in nutrigenomics. After translation, and under the impact of dietary status surrounding the primarily translated proteome, the final set of functional proteins, activated pathways, and subsequent metabolites constitutes the functional Gene Product. The gene product is only potentially functional towards a certain phenotypic outcome. The downstream end result of health status depends greatly on what nutrients are fed into the systemic machine of gene products. The functional gene product is the end-point in nutrigenomics and the starting point in nutrigenetics. It is a marker of the phenotypic outcome: expression of disease and prognosis. Phenotype may dictate the lifestyle choices available to a certain individual, including taste preferences, which are also delineated by culturally inherited customs and habits. In their turn, lifestyle choices including dietary habits determine environmental exposures. Furthermore, civilization diseases have been hidden for a long period of time due to the sociocultural inheritance of adequately evolved matching lifestyle preferences and diet choices that have been masking a biologically inherited limited gene pool. The genes being in status quo, in presence of a nutritional transition, the rates of civilization diseases are on the rise because of the loss of the protective adequacy of the diet. This highlights the presence of hidden genes, the phenotypic expression of which can be masked by a specific nutritional state, such as that corresponding to the MEDITERRANEAN DIET, as more increasingly being recommended recently in the literature. However this cannot be answered if sequence variations and specific SNPs affecting nutritional needs are not tested for in the specific populations.

- The genome sequence of DNA base pairs dictates the primary genetic profile and hence gene function. DNA sequence variants—gene variants and single nucleotide polymorphisms (SNPs)—designate an alteration of gene structure with or without functional changes that might or might not LEAD to a complex gene-function relationship depicted in different diseases [27]. The epigenome, an emerging concept in genetic research, is a set of nongenetic factors, such as diet, that change the expressed gene outcome without affecting the structure of the DNA per se.
- Nutrigenomics considers environmental factors of alimentary source that may disrupt the DNA sequence in peptide-coding and in promoter regions, affecting the gene product. Other environmental nutrigenomic factors include abundance of macro and micronutrient components of the diet, presence of other BIOACTIVE food elements, and caloric content. Under-

or overnutrition in the maternal environment sets epigenetic programming mechanisms via energetic control of function and oxidation. Through regulation of many biological functions including mitochondrial activity, cellular stress, inflammation, and telomere shortening, the dietary signature starts when epigenetic mechanisms induce or limit the risk to disease [28]. Possible levels of expression of a certain gene lie in a range of disease susceptibility that is determined by epigenetic mechanisms. These mechanisms are dictated by the functional profile of the cell, which obeys its nutritional state and reflects the nutritional environment.

- Although sustainability of epigenetic programming along life span is not well understood [28], two temporally distinct profiles may be distinguished. First, the basal epigenome is determined early on in life. Depending on the basal expressivity of DNA, it behaves like a permanently edited version of the genome. Accordingly, increasing evidence of trans-generational inheritance of epigenetics was found in mice [29] through the effect of grandmaternal nutrition on grandchildren during gamete stage, throughout the mother's fetal stage [30]. Second, later in life, similar mechanisms affect gene expressivity in response to temporary environmental factors, resulting in a short-lived epigenetic profile. These changes are mainly due to interference of nutrients and BIOACTIVE food components with transcription factor conformations [26]. This signature serves as a means for the organism to receive information about its nutritional environment in order for the cells to execute appropriate modifications on the profile of expressed genes [31]. The nutrigenomic signature is not well studied in humans yet; however obvious importance is due to its impact on gene expression, chronic diseases susceptibility, and health status of future generations [31–33].
- Regulators of posttranscriptional modifications affect alternative RNA splicing which gives rise to different mature mRNA isoforms. Alternative splicing is as highly prevalent as in 35 to 59% of human genes [34]. Post-trancriptional regulators, such as microRNAs and their coacting and counteracting proteins, are part of the RNA and protein pools [35]. They are hence influenced by epigenetic and metabolic factors as well [26, 36].

- Proteomics is the study of the protein pool in the organism, as an integral part of the cellular function. On the other hand, the metabolome designates the structure, the localization, the post-translational modifications, and the functions of proteins and metabolites along with their interactions in the organism [37]. It defines the current metabolic state and active intracellular pathways in the organism (Figure 5). The functional gene products comprise all the potentially functional molecules and pathways, whether currently active or not, that result from a certain genome-epigenome combination, LEADING to a certain range of possible phenotypic outcomes, rather than a clearly defined health status.
- Nutrigenetics is a quite different approach that emerged when dietary interventions were able to successfully alter the course of certain diseases. The basic principle considers how the same dietary environment can result in different phenotypic outcomes of health or disease in metabolizers with different functional gene-products or programmed phenotypes [38]. The concept is similar to how individuals possess different phenotypes as drug-metabolizers. The study of genetic variations affecting nutrient metabolism, from digestion to detoxification, can decipher ambiguities in the diet-disease relationship [38]. However, the challenge lies in the ability of researchers to describe the processes through which the dietary environment imposes itself to precipitate metabolic disorders.
- Finally, hypotheses of Thrifty Profile, namely thrifty genes and thrifty phenotypes, offer explanations for etiology, predisposition, and rising prevalence of DM and obesity. Early life dietary habits foretell the basal appetite control and cellular nutritional needs through psychological and molecular habituations [28]. Thrifty genes that enable survival during periods of food shortages may have been conserved over generations under the selection pressure of under-nutrition [39]. Thrifty phenotypes may be due to early nutritional challenges that enhance nutrients-saving mechanisms in the growing individual, leading to excessive storage later on and increased risk of metabolic disorders [40]. Both models have not gathered enough evidence apiece; however combined they provide a fertile base for further nutritional genomic research.

An example of a phenotype that has evolved accordingly is taste preferences and ability to digest, absorb, and appropriately respond to nutrients [41]. Genes for taste receptors, among other proteins that handle the metabolism of different nutrients, have been extensively studied. In an extensive review by Garcia-Bailo et al., an important aspect of the dietary signature is addressed: the genetic variations that affect dietary habits and food choices, with an emphasis on their effects on the nutritional environment and the health outcome [41].

Given the rise in multifactorial diseases, nutrigenetics started to involve public health research, hinting at personalized dietary recommendations for prevention of civilization diseases many years before clinical manifestations arise [31]. Adequacy of the general dietary recommendations to the ancient nature of our genes is becoming increasingly dubious. The human genome, as we know it, was sculpted throughout 2 million years of evolution under the diets of our hunters-gatherers ancestors [42]. Later on, the available food choices changed since the introduction of agriculture, but too rapidly for the ancestral stone-age genome to keep up with. This fast nutritional transition revealed evolutionary origins of obesity and diabetes among other civilization epidemics [43]. The experiment-based advancement of dietary recommendations during the past 25 years showed a convergence towards what looks more like a Paleolithic hunter-gatherer diet [44]. Despite low compliance to recommended diets and increasing industrialization of actual dietary habits, personalized and ancestral dietary recommendations still seem promising.

# NUTRITIONAL GENOMICS IN THE MENA

Diet, genetics, and disease are linked in many ways as could be shown in Figure 5. The MENA is a region that has been witnessing simultaneously a dietary change and a worsening prevalence of chronic diseases. Because of this, nutritional genomics research in such a region can improve our understanding of this rapid change in disease prevalence and shed light on the genomic effects of this dietary transition in the region. Nutritional genomics research in the MENA is minimal. To our knowledge, this is the first review on the topic in the region. We aim to collate studies in MENA countries that discuss any aspect of the dietary signature that we discussed in Figure 5. We approach that using examples of common diseases with rising prevalence in the MENA. The paper discusses two categories of diseases: (1) civilization disorders of metabolism (cardiovascular diseases and metabolic risk factors), and (2) micronutrient deficiencies (MNDs).

We also look at other populations where nutritional genomics research in these disease categories was done and discuss how it applies to our region with recommendations for future research on MENA populations.

# DIET-RELATED CIVILIZATION DISORDERS OF METABOLISM IN THE MENA REGION

Populations of the MENA belong to a unique genetic pool because of the mixture of ethnicities with horizontal mixing of populations throughout history, the high rate of consanguineous marriages within subpopulations, and the geography of the states making up the region (Figure 1). Nevertheless, wide prospective population studies on the effects of polymorphisms on such disorders of metabolism in the MENA are still lacking [25]. Based on literature reports on other populations, a large set of genes and DNA sequence variants are potentially culpable of the rise of metabolic disorders under the effect of industrialized diets. In this part of the paper, we present numerous polymorphisms that predispose to metabolic disorders including T2DM, obesity, dyslipidemias, atherosclerosis, cardiovascular events, and hypertension. The civilization disorders have common pathophysiologies and risk factors of metabolism and, since interrelated and comorbid, will accordingly be treated as one major health outcome in the following discussion about genetic entities common to the different disorders, under the effect of diet.

The Brain-Derived Neurotrophic Factor (BDNF) is important for energy balance in mice and for regulation of stress response in humans (OMIM 113505). Polymorphisms in this gene are associated with obesity and all subtypes of psychological eating disorders in Europeans (NCBI 627). Recently, three-way association was identified between hoarding behavior of obsessive-compulsive disorder, obesity, and the Val/Val genotype ofBDNF in the Valine (Val) to Methionine (Met) amino acid change at position 66 (Val66Met) in Caucasians [45]. The suggested evolutionary mechanism for this complex relationship between gene, psychopathology, and body weight is the conservation of a thrifty gene, once an old survival strategy.

Control of fetal appetite was recently shown to be a function of the Fat Mass- and Obesity-Associated (FTO) gene expression [28] which codes a nuclear oxygenase that affects tissue lipid metabolism [46] (NCBI 79068) and depends on energy balance during development [28]. A strong relation links FTO SNPs to higher risk of obesity and T2DM in many international studies (OMIM 610966) [47, 48].

Transcription Factor 7-Like 2 (TCF7L2) gene codes a transcription factor involved in blood glucose homeostasis. The rs7903146 variant association to T2DM varies greatly over ethnicities (OMIM 602228). However in Palestinians, this SNP (114758349C > T) increases the risk for T2DM, and homozygotes are affected at younger age [49].

Calpain 10 (CAPN10), which codes a calcium-dependent cysteine protease, is being increasingly studied for its role in T2DM (NCBI 11132). SNP-44 of CAPN10 has significant association with T2DM and total cholesterol in Gaza [50], while only UCSNP-19 SNP and haplotype-111 are proven to be high risks for T2DM in Tunisia [51].

The association of T2DM with polymorphisms of the Angiotensin Converting Enzyme (ACE) and theMethylene Tetrahydrofolate Reductase (MTHFR) is not well proven [52]. However data in Tunisians suggest synergistic action of the ACE Insertion/Deletion (I/D) dimorphism with the MTHFR C677T SNP on risk of T2DM [52]. Fairly common, ACE D and MTHFR 677T alleles are, respectively, present in around 77 and 27% of Moroccans [53]. Nevertheless, ACE DD genotype in Tunisians is associated with higher ACE activity and might become a useful clinical marker for CAD risk assessment of acute myocardial infarction [52, 54]. In Lebanon, ACE D allele and age, combined, are associated with higher risk for hypertension [54]. Also, Lebanese with MTHFR C677T turned out to be more susceptible to diabetic nephropathy than Bahrainis with the same SNP [55]. The SNP cannot constitute an independent risk factor in Arabs [56]. Its effect is presumably due to high homocysteine levels and hence must be evaluated depending on dietary and ethnic backgrounds [55].

In genes encoding the G protein-coupled Beta-2- and 3-Adrenergic Receptors (ADRB2, ADRB3), evolutionary selection of specific alleles exists in Africans, Asians, and Europeans. ADRB2 Glu27 and Gln27 are, respectively, factors of exercise-dependent obesity risk and metabolic syndrome susceptibility (OMIM 109690). Glu/Glu and Glu/Gln can independently predict severe Coronary Artery Disease (CAD) in Saudi Arabs [57]. However ADRB3 Trp64Arg SNP is a CAD predictor only in presence of other risk factors in Arabs, but not an independent one [57]. ADRB3 is mainly located in adipose tissues causing easier weight gain and earlier T2DM onset in Trp64Arg individuals in several populations [58] (NCBI 155) (OMIM 109691).

Peroxisome Proliferator-Activated Receptor Gamma (PPARG) genes encode nuclear receptors and regulators of adipocyte differentiation and possibly lipid metabolism and insulin sensitivity (OMIM 601487). Pro12Ala isoform of PPARG2 seems to activate transcription less effectively and carry less morbidity. Carriers of a Pro12Ala polymorphism may have a weaker BMI correlation to amount of dietary fat when compared to Pro homozygotes [59], while response to quality of dietary fat is greater in terms of BMI, lipid profile, and fasting insulin levels [60, 61]. However these associations were not found for many of the studied populations (OMIM 601487).

Apolipoproteins (APOs) are involved in lipid metabolism. APOE polymorphisms have been heavily studied. In the APOE G219T SNP, TT individuals have prolonged postprandial lipemia [62]. Apo E has three major isoforms, E2, E3, or E4. APOE E4 individuals may be protected effectively by lower dietary fat intake [63] while non-E4 individuals have minimal to no benefit from dietary intervention on lipid profile [64]. In Iranians, E2 allele was associated with lower total cholesterol levels [65]. However, despite correlation between APOE2 and LDL subfraction profiles in healthy Arabs, no similar association was found in Arabs with CAD [66]. APOEE2, E3, and E4 carriers constitute approximately 11, 79, and 10% of Moroccans, respectively [53].

Moreover, mutations in Lipoprotein Lipase (LPL), which is crucial for receptor-mediated lipoprotein uptake, drastically affect lipoprotein metabolism disorders (NCBI 4023). In Saudi Arab population however, lack of association between LPL polymorphisms and CAD was noticed [56]. Strong evidence exists for Hepatic Lipase (LIPC) C514T homozygotes. They have more atherogenic lipid profile in response to dietary fat in addition to impaired adaptation to higher animal fat with higher cardiovascular diseases risk [67, 68].

Finally, the Paraoxonase (PON1) gene encodes for an anti-atherosclerotic esterase which capacitates high-density lipoproteins to prevent lipoprotein oxidation. Gln192Arg and Leu55Met are two common polymorphisms of PON1 that modulate PON1 activity in the serum, which predicts the architecture of apolipoprotein, lipoprotein, and lipid levels [69]. In the late 1990s, PON1 status, including genotype and serum activity levels, has been proven to predict cardiovascular risk much better than genotype alone [70]. However, in Turkish subjects, there was no consistent association between the polymorphisms and the lipid levels [71]. An individual's polymorphism might be suggestive of a high risk while his dietary signature is making the actual PON1 activity favorable, that is, low risk. This interaction between diet and genes can hinder the significance of genetic screening, and enhance the relevance of proteomics and metabolomics. The lesson learned from the PON1 role in cardiovascular disease is of utmost relevance. Functional genomic analysis is required for adequate risk assessment; an individual may be screened for all known polymorphisms of PON1, but still not be assigned a risk category for cardiovascular disease [72].

Discrepancies between genotype and function impose limitations on genetic screening. Similarly for most of the polymorphisms presented previously, the degree to which genetic screening can be helpful in decision-making is controversial. More activity correlation studies are needed to examine the "penetrance" of polymorphisms. Also, insufficient nutrigenomic and proteomic evidence may be misleading [117]. Hence, further multidisciplinary studies, with coordination between laboratories, will be needed to decide which gene/polymorphism would be worth screening in a particular population.

Further multidisciplinary nutritional genomics research is needed for more specific targeted individualized advising and therapy. However, given the current lack of adequate understanding of the genetic etiologies of civilization diseases and wide-scale regional genetic screening studies, reversal of dietary changes is rendered the simplest available measure to control the metabolic epidemic of civilization diseases in the MENA.

# MICRONUTRIENT DEFICIENCIES (MNDS) IN THE MENA REGION

Micronutrients (vitamins and minerals) are required throughout life, in minute amounts in the human body, to function as cofactors of enzymes or as structural components of proteins, or to maintain genome stability, among other physiological roles [118]. Both their excess and deficiency may cause DNA damage, alter growth and development, contribute to a wide array of chronic diseases, and jeopardize health [118]. MNDs are highly prevalent in MENA countries as was established earlier. Deficiencies in iodine, iron, and vitamin A are very important MNDs in terms of prevalence and potential threat to public health worldwide; however relevant gene-diet interaction has not been sufficiently studied in the MENA. This section will thus be restricted to the following MNDs of particular interest in diet-genetics-disease interaction: vitamin D, calcium, iron, folate, and vitamins C, E, B6, and B12.

## Vitamin D Deficiency (VDD)

Vitamin D is a fat-soluble vitamin, with two forms, one present in a narrow range of foods (D2) and another formed under the skin when exposed to the ultraviolet B (UVB) light fraction of sunlight (D3); both are activated by the liver and kidneys [119] (Figure 6). Prolonged VDD can result in rickets in young children and osteoporosis and fractures in adults [120]. Recently, low vitamin D levels have been associated with increased risk of hypertension, cardiovascular diseases [121], cancer [122], diabetes [123], musculoskeletal and immunity disorders, and infectious diseases [124].

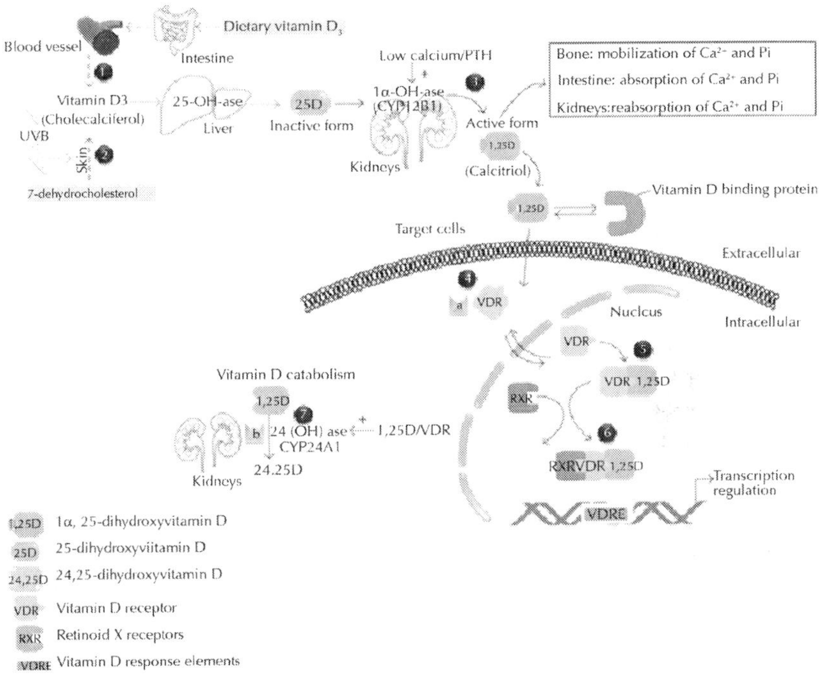


**Figure 6:** Vitamin D pathway and sites of interaction with dietary factors. Cutaneous or dietary vitamin D is hydroxylated in the liver to form 25-hydroxyvitamin D (1,2) and in the kidney to form 1 ,25-dihydroxyvitamin D (3). 1 ,25-dihydroxyvitamin D binds to VDR (4), the 1 ,25-dihydroxyvitamin D ligand promotes VDR-RXR heterodimerization (5), and the complex binds to VDRE to mediate transcriptional regulation of target genes (6). The concept of gene-diet interaction is described in the vitamin D pathway by the different polymorphisms in the VDR gene (a) and the dietary regulation of CYP24A1 enzyme (b).

Despite the sunny climate, the MENA has a highly prevalent VDD across all age groups, with the highest rate of rickets worldwide [22]. The main reasons are limited sun exposure and low dietary vitamin D intake, along with frequent pregnancies, short breastfeeding periods [125], skin pigmentation [126], body mass index [127], religious practices [128], and educational levels [129].

In addition to nutritional and social factors of VDD, genetic factors also play an important role and are depicted on the metabolic pathway of vitamin D shown in Figure 6. Genetic variations predisposing to VDD are related to Vitamin D Receptor (VDR) polymorphisms at intron 8 (BsmI) and exon 2 (FokI) [130]. The Fok1 polymorphism (C>T) in the translation initiation site creates an upstream initiation codon and a three amino acids longer molecule in the f allele compared to the F allele [131] which gives a more transcriptionally active VDR [132] LEADING to the tolerance to low vitamin D levels observed in Egyptian FF homozygotes [130]. This suggests possible evolutionary adaption to dietary intake or lifestyle changes. Only FF homozygote children have increased calcium absorption and bone mineral density [131]. Conversely, the decreased calcium absorption linked to the f allele was correlated with an increase in colon cancer risk only when calcium dietary intake is low [133]. FF genotype seems hence more advantageous than ff genotype. Paradoxically, FF (shorter VDR) was correlated with rickets unlike ff (longer VDR) in Turks and Egyptians. Thus, further studies are needed to understand the complex genetics and risks of rickets. VDR B allele also predisposes to VDD since Egyptian B homozygotes had severe rickets [130]. In other studies in the Middle East, high vitamin D doses were needed to treat patients with rickets [134]. Thus, the unexpected high prevalence of VDD in the MENA could be linked to VDR polymorphisms.

Moreover, a G>A polymorphism at position −3731 of the cdx-2 (Caudal-Type Homeobox Transcription Factor) binding element on the VDR gene promoter is another genetic variant of VDR that affects calcium absorption. The Cdx2 promoter A allele (cdx-A) binds cdx2 more strongly and has a greater transcriptional activity compared to the cdx-G allele. Thus, the A allele may increase intestinal VDR expression, subsequently enhancing calcium intestinal absorption and increasing bone mineral density. The differential expression ofVDR shows how genetic differences influence the body response to nutrients [135].

BIOACTIVE food components may exert an effect on gene expression and enzyme activity, subsequently decreasing disease risk. For example, 4′,5,7-Trihydroxyisoflavone (genistein), a soy component, is a genome-protective nutrient. It inhibits the activity of CYP24A1 (Figure 6) and thus 1,25$(OH)_2$D degradation, increasing VDR stability and the half-life and biological effects of vitamin D [136]. Folate can also inhibit this activity by increasing methylation of the promoter of CYP24A1. Also, addition of vitamin D and calcium to the western diet significantly decreases the incidence of colon cancer [133].

## Iron Deficiency

Iron (Fe) is an essential mineral needed in small amounts mainly for the production of hemoglobin and utilization of oxygen among other vital functions. Iron deficiency is a common MND mostly caused by low intake of iron, blood loss, and parasitic infections [137]. Iron absorption is enhanced by vitamin C, low pH, and heme iron and hindered by bioactive vegetables components (polyphones, tannins, phytates) and calcium [138].

Genetic factors that contribute to iron deficiency, in addition to the dietary intake, are underscored by the ability of many individuals to maintain normal iron levels despite low iron dietary intake. Nutritional iron deficiency and genetic iron deficiency have been experimentally distinguished in mice. Hephaestin (Heph) is a multicopper oxidase that allows iron basolateral surface export. Sex-linked anemia (Sla) mice bearing a deletion in Heph gene compared to control mice showed different responses to diet; Sla mice had duodenal iron accumulation and low plasma iron [139]. Additionally, mutations in the genes of human hemochromatosis protein (HFE) and its interacting protein beta-2 microglobulin (B2M), which play an important role in iron metabolism, cause murine iron deficiency [140].

## Folate Deficiency

Folic acid is a water soluble B vitamin of exclusive dietary origin. It provides the one-carbon metabolism with its main coenzyme form, tetrahydrofolate (THF). A key enzyme herein, methylenetetrahydrofolate reductase (MTHFR), catalyzes vitamin B12-dependent conversion of homocysteine to methionine, a precursor of S-adenosylmethionine (SAM), a methyl donor to DNA [141]. Thus, folate deficiency results in hyperhomocysteinemia (HHC), a risk factor for CAD [142].

MTHFR C677T is prevalent in 4% of Pakistanis [32], 2% of Yemenite Jews, 10% of Muslim Arab Israelis, and 11% of Lebanese [151]. It LEADS to a thermolabile MTHFR, which precipitates HHC in low folate states. This gene-environment combination is a risk factor for cardiovascular diseases [152], neural tube defects, and other chronic diseases [153].

C677T is a genetic variation that affects individual dietary requirements because it makes prevention of folate-related diseases that require higher folate intake. Korean C homozygotes develop HHC only with low folate levels, while T homozygotes have HHC even with normal folate levels [154]. Surprisingly however, the hyperhomocysteinemia, low folate levels, and increased cardiovascular risk observed in Indian Asians compared to European whites were not attributed to the MTHFR 677T variant [155]. Studies specific to the MENA are therefore needed because of possible variations from other regions.

The example of MTHFR and folate also underscores the genome-epigenome interplay. The TT genotype alters DNA methylation and gene expression in peripheral blood mononuclear cells only in folate deficient patients [156]. Paradoxically, the same polymorphism is inversely associated with and hence has a protective role against colorectal cancer (CRC). CRC risk decreases with adequate methionine intake, which leads to an increased formation of SAM and a negative feedback inhibition of MTHFR activity. However, the protective roles of MTHFR mutation and methionine dietary intake require an adequate dietary folate intake [157]. These findings suggest that inadequate folate intake puts carriers of particular genetic variants at higher risk of cancer. Thus personalized dietary interventions might be beneficial in reducing cancer risks.

Serum vitamin B12 deficiency is highly prevalent in Iran [158]. Thus, low folate and vitamin B12 levels can be inversely linked to the hyperhomocysteinemia observed in this population. Importantly, high prevalence of CAD was observed in a Turkish population with low plasma folate despite low plasma cholesterol concentrations [159]. It remains to determine the link between hyperhomocysteinemia, low folate intake, CAD risk, and MTHFR variants, to explain this prevalence in Turkey and possibly in other countries of the MENA region. This would add folate supplementation as a possible treatment for hyperhomocysteinemia. Other studies have also reported the possible contribution of the level of pyridoxal phosphate (PLP or vitamin B6) to hyperhomocysteinemia and vascular disease; low PLP levels were observed in individuals with the homozygote TT genotype compared to healthy individuals from some countries of the MENA [160, 161]. These data corroborate the various dietary signatures on the specific genetic profile and show that other mechanisms, enzymes, and vitamins should be examined as well.

## Vitamin B12 Deficiency

Like folate deficiency, vitamin B12 (cobalamin) deficiency can affect establishment of the disease depending on the genetic background. A common genetic variant is detected in Methionine Synthase Reductase (MTRR), an important enzyme for maintaining Methionine Synthase in its active state. The polymorphism is an A66G substitution resulting in an Ile22Met residue. The homozygous genotype was associated with an increased risk of neural tube defects (NTDs) when combined with low vitamin B12 levels. Vitamin B12 deficiency is highly prevalent among Iranian women of childbearing age [158]. However polymorphisms in both MTHFR andMTRR increase NTDs risk [162].

## Vitamins C and E Deficiencies

As antioxidants, vitamins C and E have an important function in the diet-gene interaction. Glutathione S-transferases (GSTs) transfer glutathione to different substrates. A common deletion of GSTM1 gene, a deletion polymorphism in GSTT1, and an A313G polymorphism of GSTP1 result, respectively, in a nonfunctional genotype, loss of enzyme activity, and altered activity of the GST isoforms. The GST enzymes were found protective against serum ascorbic acid deficiency when vitamin C consumption is low, since GST null genotypes with low vitamin C intake had an increased serum ascorbic acid deficiency risk [163]. The Hp1 and Hp2 polymorphisms in the hemoglobin-binding protein haptoglobin (Hp) were also studied in vitamin C deficiency. Unlike Hp1 carriers, Hp2 homozygotes had lowest serum vitamin C concentrations when dietary vitamin C intake is insufficient. Thus, Hp1 has a greater antioxidant capacity preventing hemoglobin-iron-related vitamin C oxidation and depletion [164].

A protective role for vitamin E against atherosclerosis, cancer, and neurodegenerative diseases has also been reported. Polymorphisms in the proteins involved in vitamin E metabolism lead to differential vitamin E uptake and response among individuals, and subsequently different disease risk [165]. Dietary vitamin E intake also influences the body mass index (BMI) and risk of obesity via modifying genetic variants of SIRT1 (sirtuin protein family of nicotinamide-adenine-dinucleotide- (NAD+)-dependent histone deacetylases) [166].

Well-studied gene-diet interactions are also critical in the pathophysiology of cancer. Given the evidence that VDR polymorphisms, MTHFR genotype, and DNA methylation in a low calcium or folate intake are associated with an increased cancer risk, the dietary signature greatly influences carcinogenesis. Significant dietary factors include antioxidants such as vitamin C, carotenoids, lycopene, tocopherols (vitamin E), and many other micronutrients present in fruits and vegetables.

Being one of the LEADING causes of death worldwide as well as in the MENA, cancer has been extensively studied, and the diet-genetics-cancer interaction is currently being thoroughly investigated for each and every one of the involved micronutrients. So, the effect of the diet-genetics interaction on carcinogenesis will not be dwelled upon in this paper.

The aforementioned studies collectively depict the interaction between diet, genetic variability, and disease. Genetic variants might affect gene expression patterns and epigenetic events resulting in differential body responses to diet. However, a small individualized nutritional intervention that is well studied to provide the needed concentrations of micronutrients can influence genetic variants to decrease disease risk. Thus, nutrigenetics is a tool for choosing the appropriate diet according to the individual's genetic makeup.

# A CALL FOR NUTRITIONAL GENOMICS RESEARCH IN THE MENA

All previous data support the crosstalk between diet and genome. Differential responses to dietary components among individuals are determined by genetic factors. The deleterious effects of some genotypes can be circumvented by an increased intake of particular nutrients to overcome the genetic susceptibility, which opens the horizon for personalized diet. In turn, nutrients might affect genome, gene expression, and phenotype. Although hard and complex, it is worthwhile to identify the genes that predispose individuals to chronic diseases and the nutrients that regulate their expressions to modify personal risks and to prevent, mitigate, or treat diseases. Studying on an individual basis the interactions between diet and genetics could help select appropriate diet to optimize health status. Indeed, the picture becomes more complicated when lifestyle, behavioral, and other environmental factors interfere with the diet-genetics interaction.

Of note is the unique ethnic combination of the region's native populations that make studies from other regions inapplicable to the MENA. In spite of the multiethnic origins, high rates of consanguinity in the subpopulations render the genetic pool paradoxically limited and significantly increase not only the risk of congenital abnormalities but also the susceptibility of the population to chronic diseases and genetic disorders [167]. Screening for the common polymorphisms in the MENA can give insights on their prevalence in the region or can help discover polymorphisms indigenous for the region. Such action could help alleviate the burden of chronic diseases in the MENA simply by suggesting adequate adjustments of dietary factors to hide a genetic polymorphism or to prevent DNA damage.

In some of the examples provided previously, success can be achieved, but in others, researchers ought to be more cautious. Selecting which polymorphisms are to be screened for, at the population level in the MENA, should be made after careful understanding of the effects of these variants on diet and disease. This is crucial to avoid misleading results and unnecessary costs. Unfortunately, functional studies are limited, but wide-scale screening and associations from other regions in the world could guide the decision-making process regarding screening in the MENA. At the same time, more effort and money should be invested in molecular and cellular research in nutritional genomics in order to better understand the function of the dietary signature and to more confidently guide population screening and personalized diet.

## CONCLUSIONS

(i)MENA countries are witnessing a radical change in dietary patterns from a traditional diet to a less healthy industrialized diet.(ii)Rising prevalence for civilization diseases of metabolism and micronutrient deficiencies in the MENA parallels the change in dietary habits and is mostly caused by it.(iii)Nutrigenomic factors and the dietary signature on the genome play a role in the diet-disease interactions.(iv)Genetic sequence variations, epigenetic profiles, and posttranscriptional and posttranslational modifications are some of the mechanisms that define the diet-genetics-disease relationship.(v)A large set of gene polymorphisms have been correlated with civilization diseases of metabolism, only a little of which have been studied in MENA countries.(vi)There are different mechanisms through which diet-genetics interaction affects micronutrient pathways and contributes to disease, including vitamin D and calcium, iron, folate, and vitamins C and E.(vii)Given the drastic dietary changes in the region over a short period of time, diet is the most obvious public health intervention, yet system biology and genomics research should not be underestimated. (viii)Wide-scale screening for certain gene polymorphisms in the MENA might allow for efficient intervention with personalized diet.(ix)More nutrigenomics research is needed to look at function and mechanisms of the diet-genetics-disease interaction.

# AUTHORS' CONTRIBUTION

A. C. Fahed and G. M. Nemer designed the narrative review and the different sections of the manuscript, supervised the work, and compiled the final draft. A. M. El-Hage-Sleiman and T. I. Farhat performed the literature search, designed the figures, and wrote equally different sections of the manuscript. A. K. M. El-Hage-Sleiman and T. I. Farhat contributed equally to this paper. All authors read and approved the final manuscript.

# ACKNOWLEDGMENTS

The authors would like to thank all the Nemer Lab team members for their support during the preparation of this work. They would also like to thank Dr. Amy Zenger from the Writing Center at the American University of Beirut for editing the manuscript.

# REFERENCES

1. D. Grigg, "Food consumption in the Mediterranean region," Tijdschrift voor Economische en Sociale Geografie, vol. 90, no. 4, pp. 391–409, 1999.
2. A. O. Musaiger, "Diet and prevention of coronary heart disease in the Arab Middle East countries,"Medical Principles and Practice, vol. 11, supplement 2, pp. 9–16, 2002.
3. A. Belal, "Nutrition-related chronic diseases Epidemic in UAE: can we stand to STOP it?" Sudanese Journal of Public Health, vol. 4, pp. 383–392, 2009.
4. A. Alwan, "Noncommunicable diseases: a major challenge to public health in the Region," Eastern Mediterranean Health Journal, vol. 3, pp. 6–16, 1997.
5. K. Bagchi, "Iron deficiency anaemia—an old enemy," Eastern Mediterranean Health Journal, vol. 10, no. 6, pp. 754–760, 2004.
6. D. C. Klonoff, "The beneficial effects of a Paleolithic diet on type 2 diabetes and other risk factors for cardiovascular disease," Journal of Diabetes Science and Technology, vol. 3, no. 6, pp. 1229–1232, 2009.
7. A. Trichopoulou, T. Costacou, C. Bamia, and D. Trichopoulos, "Adherence to a MEDITERRANEAN DIET and survival in a Greek population," The New England Journal of Medicine, vol. 348, no. 26, pp. 2599–2608, 2003.
8. S. Russeau and M. Batal, The Healthy Kitchen, Recipes from Rural Lebanon, Ibsar, Nature Conservation Center for Sustainable Futures American University of Beirut, 2009.
9. P. Mirmiran, R. Sherafat-Kazemzadeh, S. Jalali-Farahani, and F. Azizi, "Childhood obesity in the Middle East: a review," Eastern Mediterranean Health Journal, vol. 16, no. 9, pp. 1009–1017, 2010.
10. S. Yusuf, S. Reddy, S. Ôunpuu, and S. Anand, "Global burden of cardiovascular diseases part II: variations in cardiovascular disease by specific ethnic groups and geographic regions and prevention strategies," Circulation, vol. 104, no. 23, pp. 2855–2864, 2001.

11. P. Zimmet, D. Canteloube, and B. Genelle, "The prevalence of DIABETES MELLITUS and impaired glucose tolerance in Melanesians and part-Polynesians in rural New Caledonia and Ouvea (Loyalty Islands)," Diabetologia, vol. 23, no. 5, pp. 393–398, 1982.
12. International Diabetes Federation, IDF Diabetes Atlas, 1st edition, 2000, http://www.eatlas.idf.org/.
13. International Diabetes Federation, IDF Diabetes Atlas, 2nd edition, 2003, http://www.eatlas.idf.org/.
14. International Diabetes Federation, IDF Diabetes Atlas, 3rd edition, 2007, http://www.eatlas.idf.org/.
15. International Diabetes Federation, IDF Diabetes Atlas, 4th edition, 2010, http://www.eatlas.idf.org.
16. S. Wild, G. Roglic, A. Green, R. Sicree, and H. King, "Global prevalence of diabetes: estimates for the year 2000 and projections for 2030," Diabetes Care, vol. 27, no. 5, pp. 1047–1053, 2004
17. WHS, World Health Statistics Annual, 2005, http://www.who.int/whosis/whostat.
18. WHS, World Health Statistics Annual, 2011, http://www.who.int/whosis/whostat.
19. S. Yusuf, S. Reddy, S. Ôunpuu, and S. Anand, "Global burden of cardiovascular diseases. Part I: general considerations, the epidemiologic transition, risk factors, and impact of urbanization," Circulation, vol. 104, no. 22, pp. 2746–2753, 2001.
20. WHO, Draft Nutrition Strategy and Plan of Action for Countries of the Eastern Mediterranean Region 2010–2019, 2009.
21. O. M. Galal, "Micronutrient deficiency conditions in the middle east region: an overview," Public Health Reviews, vol. 28, no. 1–4, pp. 1–12, 2000.
22. A. Mithal, D. A. Wahl, J. P. Bonjour et al., "Global vitamin D status and determinants of hypovitaminosis D," Osteoporosis International, vol. 20, no. 11, pp. 1807–1820, 2009.
23. M. Moussavi, R. Heidarpour, A. Aminorroaya, Z. Pournaghshband, and M. Amini, "Prevalence of vitamin D deficiency in Isfahani high school students in 2004," Hormone Research, vol. 64, no. 3, pp. 144–148, 2005.

24. A. M. Siddiqui and H. Z. Kamfar, "Prevalence of vitamin D deficiency rickets in adolescent school girls in Western region, Saudi Arabia," Saudi Medical Journal, vol. 28, no. 3, pp. 441–444, 2007.
25. J. M. Ordovas and V. Mooser, "Nutrigenomics and nutrigenetics," Current Opinion in Lipidology, vol. 15, no. 2, pp. 101–108, 2004.
26. G. P. A. Kauwell, "Emerging concepts in nutrigenomics: a preview of what is to come," Nutrition in Clinical Practice, vol. 20, no. 1, pp. 75–87, 2005.
27. R. G. Gosden and A. P. Feinberg, "Genetics and epigenetics—nature›s pen-and-pencil set," The New England Journal of Medicine, vol. 356, no. 7, pp. 731–733, 2007.
28. S. Sebert, D. Sharkey, H. Budge, and M. E. Symonds, "The early programming of metabolic health: is epigenetic setting the missing link?" American Journal of Clinical Nutrition, vol. 94, no. 6, pp. 1953S–1958S, 2011.
29. R. A. Waterland and R. L. Jirtle, "Early nutrition, epigenetic changes at transposons and imprinted genes, and enhanced susceptibility to adult chronic diseases," Nutrition, vol. 20, no. 1, pp. 63–68, 2004.
30. S. M. Singh, B. Murphy, and R. L. O›Reilly, "Involvement of gene-diet/drug interaction in DNA methylation and its contribution to complex diseases: from cancer to schizophrenia," Clinical Genetics, vol. 64, no. 6, pp. 451–460, 2003.
31. R. Debusk, "The role of nutritional genomics in developing an optimal diet for humans," Nutrition in Clinical Practice, vol. 25, no. 6, pp. 627–633, 2010.
32. P. J. Stover, "Influence of human genetic variation on nutritional requirements," American Journal of Clinical Nutrition, vol. 83, no. 2, 2006.
33. R. L. Jirtle and M. K. Skinner, "Environmental epigenomics and disease susceptibility," Nature Reviews Genetics, vol. 8, no. 4, pp. 253–262, 2007.
34. B. Modrek and C. Lee, "A genomic view of alternative splicing," Nature Genetics, vol. 30, no. 1, pp. 13–19, 2002.
35. W. Filipowicz, S. N. Bhattacharyya, and N. Sonenberg, "Mechanisms of post-transcriptional regulation by microRNAs:

are the answers in sight?" Nature Reviews Genetics, vol. 9, no. 2, pp. 102–114, 2008.

36. C. D. Davis and J. Milner, "Frontiers in nutrigenomics, proteomics, metabolomics and cancer prevention," Mutation Research, vol. 551, no. 1-2, pp. 51–64, 2004.

37. A. Pardanani, E. D. Wieben, T. C. Spelsberg, and A. Tefferi, "Primer on medical genomics part IV: expression proteomics," Mayo Clinic Proceedings, vol. 77, no. 11, pp. 1185–1196, 2002.

38. A. El-Sohemy, "Nutrigenetics," Forum of Nutrition, vol. 60, pp. 25–30, 2007.

39. J. V. Neel, "Diabetes mellitus: a «thrifty» genotype rendered detrimental by «progress»?" American Journal of Human Genetics, vol. 14, pp. 353–362, 1962.

40. R. S. Lindsay and P. H. Bennett, "Type 2 diabetes, the thrifty phenotype—an overview," British Medical Bulletin, vol. 60, pp. 21–32, 2001.

41. B. Garcia-Bailo, C. Toguri, K. M. Eny, and A. El-Sohemy, "Genetic variation in taste and its influence on food selection," OMICS, vol. 13, no. 1, pp. 69–80, 2009.

42. S. B. Eaton and S. B. Eaton, "Paleolithic vs. modern diets—selected pathophysiological implications,"European Journal of Nutrition, vol. 39, no. 2, pp. 67–70, 2000.

43. A. M. Prentice, "Obesity in emerging nations: evolutionary origins and the impact of a rapid nutrition transition," Nestle Nutrition Workshop Series: Pediatric Program, vol. 63, pp. 47–57, 2009.

44. M. Konner and S. Boyd Eaton, "Paleolithic nutrition: twenty-five years later," Nutrition in Clinical Practice, vol. 25, no. 6, pp. 594–602, 2010.

45. K. R. Timpano, N. B. Schmidt, M. G. Wheaton, J. R. Wendland, and D. L. Murphy, "Consideration of the BDNF gene in relation to two phenotypes: hoarding and obesity," Journal of Abnormal Psychology, vol. 120, no. 3, pp. 700–707, 2011.

46. C. Church, S. Lee, E. A. L. Bagg et al., "A mouse model for the metabolic effects of the human fat mass and obesity associated FTO gene," PLoS Genetics, vol. 5, no. 8, Article ID e1000599, 2009.

47. C. Dina, D. Meyre, S. Gallina et al., "Variation in FTO contributes to childhood obesity and severe adult obesity," Nature Genetics, vol. 39, no. 6, pp. 724–726, 2007
48. T. M. Frayling, N. J. Timpson, M. N. Weedon et al., "A common variant in the FTO gene is associated with body mass index and predisposes to childhood and adult obesity," Science, vol. 316, no. 5826, pp. 889–894, 2007.
49. S. Ereqat, A. Nasereddin, S. Cauchi, K. Azmi, Z. Abdeen, and R. Amin, "Association of a common variant in TCF7L2 gene with type 2 diabetes mellitus in the Palestinian population," Acta Diabetologica, vol. 47, no. 1, pp. S195–S198, 2010.
50. M. M. Zaharna, A. A. Abed, and F. A. Sharif, "Calpain-10 gene polymorphism in type 2 diabetes mellitus patients in the gaza strip," Medical Principles and Practice, vol. 19, no. 6, pp. 457–462, 2010.
51. I. Ezzidi, N. Mtiraoui, R. Nemr et al., "Variants within the calpain-10 gene and relationships with type 2 diabetes (T2DM) and T2DM-related traits among Tunisian Arabs," Diabetes and Metabolism, vol. 36, no. 5, pp. 357–362, 2010.
52. S. Mehri, B. Baudin, S. Mahjoub et al., "Angiotensin-converting enzyme insertion/deletion gene polymorphism in a Tunisian healthy and acute myocardial infarction population," Genetic Testing and Molecular Biomarkers, vol. 14, no. 1, pp. 85–91, 2010.
53. T. P. They-They, K. Hamzi, M. T. Moutawafik, H. Bellayou, M. El Messal, and S. Nadifi, "Prevalence of angiotensin-converting enzyme, methylenetetrahydrofolate reductase, Factor v Leiden, prothrombin and apolipoprotein e gene polymorphisms in Morocco," Annals of Human Biology, vol. 37, no. 6, pp. 767–777, 2010.
54. M. Akra-Ismail, R. F. Makki, H. N. Chmaisse, A. Kazma, and N. K. Zgheib, "Association between angiotensin-converting enzyme insertion/deletion genetic polymorphism and hypertension in a sample of Lebanese patients," Genetic Testing and Molecular Biomarkers, vol. 14, no. 6, pp. 787–792, 2010.
55. R. Nemr, R. A. Salman, L. H. Jawad, E. A. Juma, S. H. Keleshian, and W. Y. Almawi, "Differential contribution of MTHFR C677T variant to the risk of diabetic nephropathy in Lebanese and

Bahraini Arabs," Clinical Chemistry and Laboratory Medicine, vol. 48, no. 8, pp. 1091–1094, 2010.

56. K. K. Abu-Amero, C. A. Wyngaard, O. M. Al-Boudari, M. Kambouris, and N. Dzimiri, "Lack of association of lipoprotein lipase gene polymorphisms with coronary artery disease in the Saudi Arab population," Archives of Pathology and Laboratory Medicine, vol. 127, no. 5, pp. 597–600, 2003.

57. K. K. Abu-Amero, O. M. Al-Boudari, G. H. Mohamed, and N. Dzimiri, "The Glu27 genotypes of the Beta2-adrenergic receptor are predictors for severe coronary artery disease," BMC Medical Genetics, vol. 7, article 31, 2006.

58. K. Shiwaku, A. Nogi, E. Anuurad et al., "Difficulty in losing weight by behavioral intervention for women with Trp64Arg polymorphism of the 3-adrenergic receptor gene," International Journal of Obesity, vol. 27, no. 9, pp. 1028–1036, 2003.

59. A. Memisoglu, F. B. Hu, S. E. Hankinson et al., "Interaction between a peroxisome proliferator-activated receptor gene polymorphism and dietary fat intake in relation to body mass," Human Molecular Genetics, vol. 12, no. 22, pp. 2923–2929, 2003

60. J. Luan, P. O. Browne, A. H. Harding et al., "Evidence for gene-nutrient interaction at the PPAR locus,"Diabetes, vol. 50, no. 3, pp. 686–689, 2001.

61. V. Lindi, U. Schwab, A. Louheranta et al., "Impact of the Pro12Ala polymorphism of the PPAR- 2 gene on serum triacylglycerol response to n-3 fatty acid supplementation," Molecular Genetics and Metabolism, vol. 79, no. 1, pp. 52–60, 2003.

62. J. A. Moreno, J. López-Miranda, C. Marín et al., "The influence of the apolipoprotein E gene promoter (-219G/T) polymorphism on postprandial lipoprotein metabolism in young normolipemic males,"Journal of Lipid Research, vol. 44, no. 11, pp. 2059–2064, 2003.

63. G. J. Petot, F. Traore, S. M. Debanne, A. J. Lerner, K. A. Smyth, and R. P. Friedland, "Interactions of apolipoprotein E genotype and dietary fat intake of healthy older persons during mid-adult life,"Metabolism, vol. 52, no. 3, pp. 279–281, 2003.

64. B. J. Nicklas, R. E. Ferrell, L. B. Bunyard, D. M. Berman, K. E. Dennis, and A. P. Goldberg, "Effects of apolipoprotein E

genotype on dietary-induced changes in high-density lipoprotein cholesterol in obese postmenopausal women," Metabolism, vol. 51, no. 7, pp. 853–858, 2002.

65. J. T. Bazzaz, M. Nazari, H. Nazem et al., "Apolipoprotein e gene polymorphism and total serum cholesterol level in Iranian population," Journal of Postgraduate Medicine, vol. 56, no. 3, pp. 173–175, 2010.
66. A. O. Akanji, C. G. Suresh, H. R. Fatania, R. Al-Radwan, and M. Zubaid, "Associations of apolipoprotein E polymorphism with low-density lipoprotein size and subfraction profiles in Arab patients with coronary heart disease," Metabolism, vol. 56, no. 4, pp. 484–490, 2007
67. J. M. Ordovas, D. Corella, S. Demissie et al., "Dietary fat intake determines the effect of a common polymorphism in the hepatic lipase gene promoter on high-density lipoprotein metabolism: evidence of a strong dose effect in this gene-nutrient interaction in the Framingham study," Circulation, vol. 106, no. 18, pp. 2315–2321, 2002.
68. E. S. Tai, D. Corella, M. Deurenberg-Yap et al., "Dietary fat interacts with the -514C>T polymorphism in the hepatic lipase gene promoter on plasma lipid profiles in a Multiethnic Asian population: the 1998 Singapore National Health Survey," Journal of Nutrition, vol. 133, no. 11, pp. 3399–3408, 2003.
69. L. S. Rozek, T. S. Hatsukami, R. J. Richter et al., "The correlation of paraoxonase (PON1) activity with lipid and lipoprotein levels differs with vascular disease status," Journal of Lipid Research, vol. 46, no. 9, pp. 1888–1895, 2005.
70. G. P. Jarvik, L. S. Rozek, V. H. Brophy et al., "Paraoxonase (PON1) phenotype is a better predictor of vascular disease than is PON1192 or PON155 genotype," Arteriosclerosis, Thrombosis, and Vascular Biology, vol. 20, no. 11, pp. 2441–2447, 2000.
71. B. Agachan, H. Yilmaz, Z. Karaali, and T. Isbir, "Paraoxonase 55 and 192 polymorphism and its relationship to serum paraoxonase activity and serum lipids in Turkish patients with non-insulin dependent diabetes mellitus," Cell Biochemistry and Function, vol. 22, no. 3, pp. 163–168, 2004
72. L. G. Costa, A. Vitalone, T. B. Cole, and C. E. Furlong, "Modulation of paraoxonase (PON1) activity,"Biochemical Pharmacology, vol.

69, no. 4, pp. 541–550, 2005.

73. I. Nabipour, M. Amiri, S. R. Imami et al., "The metabolic syndrome and nonfatal ischemic heart disease; a population-based study," International Journal of Cardiology, vol. 118, no. 1, pp. 48–53, 2007.
74. M. Nsour, Z. Mahfoud, M. N. Kanaan, and A. Balbeissi, "Prevalence and predictors of non-fatal myocardial infarction in Jordan," Eastern Mediterranean Health Journal, vol. 14, no. 4, pp. 818–830, 2008.
75. M. M. Al-Nozha, M. R. Arafah, Y. Y. Al-Mazrou et al., "Coronary artery disease in Saudi Arabia," Saudi Medical Journal, vol. 25, no. 9, pp. 1165–1171, 2004.
76. H. Ben Romdhane, R. Khaldi, A. Oueslati, and H. Skhiri, "Transition épidémiologique et transition alimentaire et nutritionnelle en Tunisie," Options Méditerranéennes B, vol. 41, 2002.
77. A. Al-Jishi and P. Mohan, "Profile of stroke in Bahrain," Neurosciences, vol. 5, no. 1, pp. 30–34, 2000.
78. A. A. Ahangar, S. B. A. Vaghefi, and M. Ramaezani, "Epidemiological evaluation of stroke in Babol, Northern Iran (2001–2003)," European Neurology, vol. 54, no. 2, pp. 93–97, 2005.
79. N. U. A. M. A. Abdul-Ghaffar, M. R. El-Sonbaty, M. S. El-Din Abdul-Baky, A. A. Marafie, and A. M. Al-Said, "Stroke in Kuwait: a three-year prospective study," Neuroepidemiology, vol. 16, no. 1, pp. 40–47, 1997.
80. K. Radhakrishnan, P. P. Ashok, R. Sridharan, and M. A. El-Mangoush, "Incidence and pattern of cerebrovascular diseases in Benghazi, Libya," Journal of Neurology Neurosurgery and Psychiatry, vol. 49, no. 5, pp. 519–523, 1986.
81. W. M. Sweileh, A. F. Sawalha, S. M. Al-Aqad, et al., "The epidemiology of stroke in northern palestine: a 1-year, hospital-based study," Journal of Stroke and Cerebrovascular Diseases, vol. 17, no. 6, pp. 406–411, 2008.
82. A. Hamad, A. Hamad, T. E. O. Sokrab, S. Momeni, B. Mesraoua, and A. Lingren, "Stroke in Qatar: a one-year, hospital-based study," Journal of Stroke and Cerebrovascular Diseases, vol. 10,

no. 5, pp. 236–241, 2001.

83. S. Al-Rajeh, E. B. Larbi, O. Bademosi et al., "Stroke register: experience from the Eastern Province of Saudi Arabia," Cerebrovascular Diseases, vol. 8, no. 2, pp. 86–89, 1998.
84. A. Yahia-Berrouiguet, M. Benyoucef, K. Meguenni, and M. Brouri, "Prevalence of cardiovascular risk factors: a survey at Tlemcen (Algeria)," Medecine des Maladies Metaboliques, vol. 3, no. 3, pp. 313–319, 2009.
85. F. I. Al-Zurba, "Latest studies clarify state of health in Bahrain," Diabetes Voice, vol. 46, pp. 28–31, 2001.
86. M. M. Ibrahim, H. Rizk, L. J. Appel et al., "Hypertension prevalence, awareness, treatment, and control in Egypt: results from the Egyptian National Hypertension Project (NHP)," Hypertension, vol. 26, no. 6, pp. 886–890, 1995.
87. O. M. Galal, "The nutrition transition in Egypt: obesity, undernutrition and the food consumption context," Public Health Nutrition, vol. 5, no. 1, pp. 141–148, 2002.
88. N. Sarraf-Zadegan, M. Boshtam, S. Mostafavi, and M. Rafiei, "Prevalence of hypertension and associated risk factors in Isfahan, Islamic Republic of Iran," Eastern Mediterranean Health Journal, vol. 5, no. 5, pp. 992–1001, 1999.
89. A. A. Haghdoost, B. Sadeghirad, and M. Rezazadehkermani, "Epidemiology and heterogeneity of hypertension in Iran: a systematic review," Archives of Iranian Medicine, vol. 11, no. 4, pp. 444–452, 2008.
90. WHO, "STEPwise surveillance. Non-communicable diseases risk factors. STEPwise data from selected countries in the Eastern Mediterranean Region, 2003–2007," 2007,http://www.emro.who.int/ncd/risk_factors.htm#physical.
91. M. Zindah, A. Belbeisi, H. Walke, and A. H. Mokdad, "Obesity and diabetes in Jordan: findings from the behavioral risk factor surveillance system, 2004," Preventing Chronic Disease, vol. 5, pp. 1–8, 2008.
92. A.-M. Sibai, O. Obeid, M. Batal, N. Adra, D. E. Khoury, and N. Hwalla, "Prevalence and correlates of metabolic syndrome in an adult Lebanese population," CVD Prevention and Control, vol. 3,

no. 2, pp. 83–90, 2008.

93. M. A. Tazi, S. Abir-Khalil, N. Chaouki et al., "Prevalence of the main cardiovascular risk factors in Morocco: results of a National Survey, 2000," Journal of Hypertension, vol. 21, no. 5, pp. 897–903, 2003.

94. A. A. Hasab, A. Jaffer, and Z. Hallaj, "Blood pressure patterns among the Omani population," Eastern Mediterranean Health Journal, vol. 5, no. 1, pp. 46–54, 1999.

95. H. F. Abdul-Rahim, G. Holmboe-Ottesen, L. C. M. Stene et al., "Obesity in a rural and an urban Palestinian West Bank population," International Journal of Obesity, vol. 27, no. 1, pp. 140–146, 2003

96. A. Bener, J. Al-Suwaidi, K. Al-Jaber, S. Al-Marri, M. H. Dagash, and I. E. Elbagi, "The prevalence of hypertension and its associated risk factors in a newly developed country," Saudi Medical Journal, vol. 25, no. 7, pp. 918–922, 2004.

97. M. M. Al-Nozha, M. Abdullah, M. R. Arafah et al., "Hypertension in Saudi Arabia," Saudi Medical Journal, vol. 28, no. 1, pp. 77–84, 2007.

98. W. Maziak, S. Rastam, F. Mzayek, K. D. Ward, T. Eissenberg, and U. Keil, "Cardiovascular health among adults in Syria: a model from developing countries," Annals of Epidemiology, vol. 17, no. 9, pp. 713–720, 2007.

99. H. M. Sonmez, O. Basak, C. Camci et al., "The epidemiology of elevated blood pressure as an estimate for hypertension in Aydin, Turkey," Journal of Human Hypertension, vol. 13, no. 6, pp. 399–404, 1999.

100. L. M. Baynouna, A. D. Revel, N. J. Nagelkerke et al., "High prevalence of the cardiovascular risk factors in Al-Ain, United Arab Emirates. An emerging health care priority," Saudi Medical Journal, vol. 29, no. 8, pp. 1173–1178, 2008.

101. A. A. Gunaid and A. M. Assabri, "Prevalence of type 2 diabetes and other cardiovascular risk factors in a semirural area in Yemen," La Revue de Santé de la Méditerranée Orientale, vol. 14, no. 1, pp. 42–56, 2008.

102. B. Motlagh, M. O›Donnell, and S. Yusuf, "Prevalence of cardiovascular risk factors in the middle east: a systematic review," European Journal of Cardiovascular Prevention and

Rehabilitation, vol. 16, no. 3, pp. 268–280, 2009.

103. A. Mehio Sibai, L. Nasreddine, A. H. Mokdad, N. Adra, M. Tabet, and N. Hwalla, "Nutrition transition and cardiovascular disease risk factors in Middle East and North Africa countries: reviewing the evidence," Annals of Nutrition and Metabolism, vol. 57, no. 3-4, pp. 193–203, 2010

104. Y. Khader, A. Bateiha, M. El-Khateeb, A. Al-Shaikh, and K. Ajlouni, "High prevalence of the metabolic syndrome among Northern Jordanians," Journal of Diabetes and Its Complications, vol. 21, no. 4, pp. 214–219, 2007.

105. I. Al Rashdan and Y. Al Nesef, "Prevalence of overweight, obesity, and metabolic syndrome among adult Kuwaitis: results from community-based national survey," Angiology, vol. 61, no. 1, pp. 42–48, 2010.

106. M. Rguibi and R. Belahsen, "Metabolic syndrome among Moroccan Sahraoui adult women," American Journal of Human Biology, vol. 16, no. 5, pp. 598–601, 2004.

107. J. A. Al-Lawati and P. Jousilahti, "Prevalence of metabolic syndrome in Oman using the International Diabetes Federation›s criteria," Saudi Medical Journal, vol. 27, no. 12, pp. 1925–1926, 2006.

108. M. Musallam, A. Bener, M. Zirie et al., "Metabolic syndrome and its components among Qatari population," International Journal of Food Safety, Nutrition and Public Health, vol. 1, no. 1, pp. 88–102, 2008.

109. A. Bener, M. Zirie, M. Musallam, Y. S. Khader, and A. O. Al-Hamaq, "Prevalence of metabolic syndrome according to adult treatment panel III and international diabetes federation criteria: a population-based study," Metabolic Syndrome and Related Disorders, vol. 7, no. 3, pp. 221–230, 2009.·

110. M. M. Al-Nozha, A. Al-Khadra, M. R. Arafah et al., "Metabolic syndrome in Saudi Arabia," Saudi Medical Journal, vol. 26, no. 12, pp. 1918–1925, 2005.

111. R. Bouguerra, L. Ben Salem, H. Alberti et al., "Prevalence of metabolic abnormalities in the Tunisian adults: a population based study," Diabetes and Metabolism, vol. 32, no. 3, pp. 215–

221, 2006.

112. M. Malik and S. A. Razig, "The prevalence of the metabolic syndrome among the multiethnic population of the United Arab Emirates: a report of a national survey," Metabolic Syndrome and Related Disorders, vol. 6, no. 3, pp. 177–186, 2008.
113. P. M. Kearney, M. Whelton, K. Reynolds, P. K. Whelton, and J. He, "Worldwide prevalence of hypertension: a systematic review," Journal of Hypertension, vol. 22, no. 1, pp. 11–19, 2004.
114. T. A. Gaziano, A. Bitton, S. Anand, S. Abrahams-Gessel, and A. Murphy, "Growing epidemic of coronary heart disease in low- and middle-income countries," Current Problems in Cardiology, vol. 35, no. 2, pp. 72–115, 2010.
115. R. M. Mabry, M. M. Reeves, E. G. Eakin, and N. Owen, "Gender differences in prevalence of the metabolic syndrome in Gulf Cooperation Council Countries: a systematic review," Diabetic Medicine, vol. 27, no. 5, pp. 593–597, 2010.
116. J. Tran, M. Mirzaei, L. Anderson, and S. R. Leeder, "The epidemiology of stroke in the Middle East and North Africa," Journal of the Neurological Sciences, vol. 295, no. 1-2, pp. 38–40, 2010.
117. G. P. Page, J. W. Edwards, S. Barnes, R. Weindruch, and D. B. Allison, "A design and statistical perspective on microarray gene expression studies in nutrition: the need for playful creativity and scientific hard-mindedness," Nutrition, vol. 19, no. 11-12, pp. 997–1000, 2003.
118. M. F. Fenech, "Dietary reference values of individual micronutrients and nutriomes for genome damage prevention: current status and a road map to the future," American Journal of Clinical Nutrition, vol. 91, no. 5, pp. 1438S–1454S, 2010.
119. S. Kimball, G. E. H. Fuleihan, and R. Vieth, "Vitamin D: a growing perspective," Critical Reviews in Clinical Laboratory Sciences, vol. 45, no. 4, pp. 339–414, 2008.
120. R. Chesney, "Metabolic bone disease," in Nelson Textbook of Pediatrics, R. E. Behrman, R. Kliegman, and H. B. Jenson, Eds., pp. 2132–2138, WB Saunders, Philadelphia, Pa, USA, 16th edition, 2000.

121. S. Wang, "Epidemiology of vitamin D in health and disease," Nutrition Research Reviews, vol. 22, no. 2, pp. 188–203, 2009.
122. C. F. Garland, F. C. Garland, E. D. Gorham et al., "The role of vitamin D in cancer prevention,"American Journal of Public Health, vol. 96, no. 2, pp. 252–261, 2006.
123. M. Janner, P. Ballinari, P. E. Mullis, and C. E. Flück, "High prevalence of vitamin D deficiency in children and adolescents with type 1 diabetes," Swiss Medical Weekly, vol. 140, article w13091, 2010.
124. J. S. Adams and M. Hewison, "Update in vitamin D," Journal of Clinical Endocrinology and Metabolism, vol. 95, no. 2, pp. 471–478, 2010.
125. S. R. Kreiter, R. P. Schwartz, H. N. Kirkman, P. A. Charlton, A. S. Calikoglu, and M. L. Davenport, "Nutritional rickets in African American breast-fed infants," Journal of Pediatrics, vol. 137, no. 2, pp. 153–157, 2000.
126. N. G. Jablonski and G. Chaplin, "The evolution of human skin coloration," Journal of Human Evolution, vol. 39, no. 1, pp. 57–106, 2000.
127. P. Lee, J. R. Greenfield, M. J. Seibel, J. A. Eisman, and J. R. Center, "Adequacy of vitamin D replacement in severe deficiency is dependent on body mass index," American Journal of Medicine, vol. 122, no. 11, pp. 1056–1060, 2009.
128. M. H. Gannagé-Yared, G. Maalouf, S. Khalife et al., "Prevalence and predictors of vitamin D inadequacy amongst Lebanese osteoporotic women," British Journal of Nutrition, vol. 101, no. 4, pp. 487–491, 2009.
129. K. Holvik, H. E. Meyer, E. Haug, and L. Brunvand, "Prevalence and predictors of vitamin D deficiency in five immigrant groups living in Oslo, Norway: the Oslo immigrant health study," European Journal of Clinical Nutrition, vol. 59, no. 1, pp. 57–63, 2005.
130. G. I. Baroncelli, A. Bereket, M. El Kholy et al., "Rickets in the Middle East: role of environment and genetic predisposition," Journal of Clinical Endocrinology and Metabolism, vol. 93, no. 5, pp. 1743–1750, 2008.
131. S. K. Ames, K. J. Ellis, S. K. Gunn, K. C. Copeland, and S. A. Abrams, "Vitamin D receptor gene Fok1 polymorphism predicts

calcium absorption and bone mineral density in children," Journal of Bone and Mineral Research, vol. 14, no. 5, pp. 740–746, 1999.

132. H. Arai, K. I. Miyamoto, Y. Taketani et al., "A vitamin D receptor gene polymorphism in the translation initiation codon: effect on protein activity and relation to bone mineral density in Japanese women,"Journal of Bone and Mineral Research, vol. 12, no. 6, pp. 915–921, 1997.
133. C. D. Davis, "Vitamin D and cancer: current dilemmas and future research needs," American Journal of Clinical Nutrition, vol. 88, no. 2, 2008.
134. M. A. Abdullah, H. S. Salhi, L. A. Bakry et al., "Adolescent rickets in Saudi Arabia: a rich and sunny country," Journal of Pediatric Endocrinology and Metabolism, vol. 15, no. 7, pp. 1017–1025, 2002.
135. H. Arai, K. I. Miyamoto, M. Yoshida et al., "The polymorphism in the caudal-related homeodomain protein Cdx-2 binding element in the human vitamin D receptor gene," Journal of Bone and Mineral Research, vol. 16, no. 7, pp. 1256–1264, 2001.
136. A. V. Krishnan, S. Swami, J. Moreno, R. B. Bhattacharyya, D. M. Peehl, and D. Feldman, "Potentiation of the growth-inhibitory effects of vitamin D in prostate cancer by genistein," Nutrition Reviews, vol. 65, no. 8, pp. S121–123, 2007.
137. A. Ali, G. Fathy, H. Fathy, and N. Abd El-Ghaffar, "Epidemiology of iron deficiency anaemia: effect of physical growth in primary school children, the importance of hookworms," International Journal of Academic Research, vol. 3, pp. 495–500, 2011.
138. W. Burke, G. Imperatore, and M. Reyes, "Iron deficiency and iron overload: effects of diet and genes,"Proceedings of the Nutrition Society, vol. 60, no. 1, pp. 73–80, 2001.
139. Chung and R. Wessling-Resnick, "Lessons learned from genetic and nutritional iron deficiencies,"Nutrition Reviews, vol. 62, no. 5, pp. 212–215, 2004.
140. J. E. Levy, L. K. Montross, and N. C. Andrews, "Genes that modify the hemochromatosis phenotype in mice," Journal of Clinical Investigation, vol. 105, no. 9, pp. 1209–1216, 2000.
141. L. B. Bailey and J. F. Gregory, "Folate metabolism and

requirements," Journal of Nutrition, vol. 129, no. 4, pp. 779–782, 1999.

142. S. W. Choi and J. B. Mason, "Folate and carcinogenesis: an integrated scheme," Journal of Nutrition, vol. 130, no. 2, pp. 129–132, 2000.

143. M. Moussavi, R. Heidarpour, A. Aminorroaya, Z. Pournaghshband, and M. Amini, "Prevalence of vitamin D deficiency in Isfahani high school students in 2004," Hormone Research, vol. 64, no. 3, pp. 144–148, 2005.

144. A. Batieha, Y. Khader, H. Jaddou et al., "Vitamin D status in Jordan: dress style and gender discrepancies," Annals of Nutrition and Metabolism, vol. 58, no. 1, pp. 10–18, 2011.

145. G. El-Hajj Fuleihan, M. Nabulsi, M. Choucair et al., "Hypovitaminosis D in healthy schoolchildren,"Pediatrics, vol. 107, article E53, 2001.

146. G. El-Hajj Fuleihan, M. Nabulsi, H. Tamim et al., "Effect of vitamin D replacement on musculoskeletal parameters in school children: a randomized controlled trial," The Journal of Clinical Endocrinology and Metabolism, vol. 91, no. 2, pp. 405–412, 2006.

147. A. Arabi, R. El Rassi, and G. El-Hajj Fuleihan, "Hypovitaminosis D in developing countries-prevalence, risk factors and outcomes," Nature Reviews Endocrinology, vol. 6, no. 10, pp. 550–561, 2010.

148. A. M. Siddiqui and H. Z. Kamfar, "Prevalence of vitamin D deficiency rickets in adolescent school girls in Western region, Saudi Arabia," Saudi Medical Journal, vol. 28, no. 3, pp. 441–444, 2007.

149. A. O. Musaiger, "Iron deficiency anaemia among children and pregnant women in the arab gulf countries: the need for action," Nutrition and Health, vol. 16, no. 3, pp. 161–171, 2002.

150. K. Bagchi, "Iron deficiency anaemia—an old enemy," La Revue de Santé de la Méditerranée Orientale, vol. 10, no. 6, pp. 754–760, 2004.

151. W. Y. Almawi, R. R. Finan, H. Tamim, J. L. Daccache, and N. Irani-Hakime, "Differences in the Frequency of the C677T Mutation in the Methylenetetrahydrofolate Reductase (MTHFR) Gene Among the Lebanese Population," American Journal of Hematology, vol.

76, no. 1, pp. 85–87, 2004.

152. A. M. T. Engbersen, D. G. Franken, G. H. J. Boers, E. M. B. Stevens, F. J. M. Trijbels, and H. J. Blom, "Thermolabile 5,10-methylenetetrahydrofolate reductase as a cause of mild hyperhomocysteinemia,"American Journal of Human Genetics, vol. 56, no. 1, pp. 142–150, 1995.
153. N. M. van der Put and H. J. Blom, "Neural tube defects and a disturbed folate dependent homocysteine metabolism," European Journal of Obstetrics Gynecology and Reproductive Biology, vol. 92, no. 1, pp. 57–61, 2000.
154. H. J. Huh, H. S. Chi, E. H. Shim, S. Jang, and C. J. Park, "Gene-nutrition interactions in coronary artery disease: correlation between the MTHFR C677T polymorphism and folate and homocysteine status in a Korean population," Thrombosis Research, vol. 117, no. 5, pp. 501–506, 2006.
155. J. C. Chambers, H. Ireland, E. Thompson et al., "Methylenetetrahydrofolate reductase 677 C→T mutation and coronary heart disease risk in UK Indian Asians," Arteriosclerosis, Thrombosis, and Vascular Biology, vol. 20, no. 11, pp. 2448–2452, 2000.
156. S. A. Ross and L. Poirier, "Proceedings of the Trans-HHS Workshop: diet, DNA methylation processes and health," Journal of Nutrition, vol. 132, no. 8, 2002.
157. J. Ma, M. J. Stampfer, E. Giovannucci et al., "Methylenetetrahydrofolate reductase polymorsphism, dietary interactions, and risk of colorectal cancer," Cancer Research, vol. 57, no. 6, pp. 1098–1102, 1997.
158. Z. Abdollahi, I. Elmadfa, A. Djazayeri et al., "Folate, vitamin B12 and homocysteine status in women of childbearing age: baseline data of folic acid wheat flour fortification in Iran," Annals of Nutrition and Metabolism, vol. 53, no. 2, pp. 143–150, 2008.
159. S. L. Tokgözoğlu, M. Alika ifoğlu, I. Ünsal et al., "Methylene tetrahydrofolate reductase genotype and the risk and extent of coronary artery disease in a population with low plasma folate," Heart, vol. 81, no. 5, pp. 518–522, 1999.
160. B. Christensen, P. Frosst, S. Lussier-Cacan et al., "Correlation of a common mutation in the methylenetetrahydrofolate reductase

gene with plasma homocysteine in patients with premature coronary artery disease," Arteriosclerosis, Thrombosis, and Vascular Biology, vol. 17, no. 3, pp. 569–573, 1997.

161. A. H. Messika, D. N. Kaluski, E. Lev et al., "Nutrigenetic impact of daily folate intake on plasma homocysteine and folate levels in patients with different methylenetetrahydrofolate reductase genotypes," European Journal of Cardiovascular Prevention and Rehabilitation, vol. 17, no. 6, pp. 701–705, 2010.

162. A. Wilson, R. Platt, Q. Wu et al., "A common variant in methionine synthase reductase combined with low cobalamin (Vitamin B12) increases risk for spina bifida," Molecular Genetics and Metabolism, vol. 67, no. 4, pp. 317–323, 1999.

163. L. E. Cahill, B. Fontaine-Bisson, and A. El-Sohemy, "Functional genetic variants of glutathione S-transferase protect against serum ascorbic acid deficiency," American Journal of Clinical Nutrition, vol. 90, no. 5, pp. 1411–1417, 2009.

164. L. E. Cahill and A. El-Sohemy, "Haptoglobin genotype modifies the association between dietary vitamin C and serum ascorbic acid deficiency," American Journal of Clinical Nutrition, vol. 92, no. 6, pp. 1494–1500, 2010.

165. J. M. Zingg, A. Azzi, and M. Meydani, "Genetic polymorphisms as determinants for disease-preventive effects of vitamin E," Nutrition Reviews, vol. 66, no. 7, pp. 406–414, 2008

166. M. C. Zillikens, J. B. J. Van Meurs, F. Rivadeneira et al., "Interactions between dietary vitamin E intake and SIRT1 genetic variation influence body mass index," American Journal of Clinical Nutrition, vol. 91, no. 5, pp. 1387–1393, 2010.

167. H. Hamamy and A. Alwan, "Genetic disorders and congenital abnormalities: strategies for reducing the burden in the Region," Eastern Mediterranean Health Journal, vol. 3, no. 1, pp. 123–132, 1997.

Chapter 2

# Role of Nutrition in the Management of Hepatic Encephalopathy in End-Stage Liver Failure

Chantal Bémeur[1,2], Paul Desjardins[1], and Roger F Butterworth[1]

[1]Neuroscience Research Unit, CHUM, Saint-Luc Hospital, University of Montreal, 1058 St-Denis Street, Montreal, QC, Canada

[2]Department of Nutrition, University of Montreal, Montreal, QC, Canada

## ABSTRACT

Malnutrition is common in patients with end-stage liver failure and hepatic encephalopathy, and is considered a significant prognostic factor affecting quality of life, outcome, and survival. The liver plays a crucial role in the regulation of nutrition by trafficking the metabolism of nutrients, their distribution and appropriate use by the body. Nutritional consequences with the potential to cause nervous system dysfunction occur in liver failure, and many factors contribute to malnutrition in hepatic failure. Among them are inadequate dietary intake, malabsorption, increased protein losses, hypermetabolism, insulin resistance, gastrointestinal bleeding, ascites, inflammation/ infection, and hyponatremia. Patients at risk of malnutrition are relatively difficult to identify since liver disease may interfere with biomarkers of malnutrition. The supplementation of the diet with amino acids, antioxidants, vitamins as well as probiotics in addition to meeting energy and protein requirements may improve nutritional status, liver function, and hepatic encephalopathy in patients with end-stage liver failure.

# INTRODUCTION

Malnutrition is a common complication of end-stage liver failure (cirrhosis) and is an important prognostic indicator of clinical outcome (survival rate, length of hospital stay, posttransplantation morbidity, and quality of life) in patients with cirrhosis. Several studies have evaluated nutritional status in patients with liver cirrhosis of different etiologies and varying degrees of liver insufficiency [1, 2] leading to a consensus of opinion that malnutrition is recognizable in all forms of cirrhosis [3] and that the prevalence of malnutrition in cirrhosis has been estimated to range from 65%–100% [4, 5]. The causes of malnutrition in liver disease are complex and multifactorial.

The present paper reviews the role of nutrition in relation to the management of hepatic encephalopathy (HE), a major neuropsychiatric complication of end-stage liver failure. Nutritional consequences of liver failure with the potential to cause central nervous system dysfunction are reviewed. In particular, the roles of dietary protein (animal versus vegetable), branched-chain amino acids, dietary fibre, probiotics, vitamins and antioxidants, minerals (zinc, magnesium) as well as L-carnitine in relation to HE are discussed. An update of the impact of nutritional supplementation on the management of HE is included.

# MALNUTRITION IN LIVER DISEASE

The functional integrity of the liver is essential for nutrient supply (carbohydrates, fat, and proteins), and the liver plays a fundamental role in intermediary metabolism. For example, the liver regulates the synthesis, storage, and breakdown of glycogen, and hepatocytes express enzymes that enable them to synthesize glucose from various precursors such as amino acids, pyruvate, and lactate (gluconeogenesis). In addition, the liver is a major site of fatty acid breakdown and triglyceride synthesis. The breakdown of fatty acids provides an alternative source of energy when glucose is limited during, for example, fasting or starvation. The liver also plays a crucial role in the synthesis and degradation of protein. Protein synthesis by the liver is influenced by the nutritional state, as well as by hormones and alcohol.

The liver plays a central role in the regulation of nutrition by trafficking the metabolism of nutrients, and many factors disrupt this metabolic balance in end-stage liver failure. Consequently, when the liver fails, numerous nutritional problems occur (Table 1). Several factors contribute to malnutrition in liver failure including inadequate dietary intake of nutrients, reduction in their synthesis or absorption (diminished protein synthesis, malabsorption), increased protein loss, disturbances in substrate utilization, a hypermetabolic state as well as increased energy-protein expenditure and requirements. Because of decreased glycogen stores and gluconeogenesis [6], energy metabolism may shift from carbohydrate to fat oxidation [7] while insulin resistance may also develop. Consequently, liver cirrhosis frequently results in a catabolic state resulting in a lack of essential nutrients.

**Table 1:** Metabolic alterations leading to malnutrition in end-stage liver failure

| Protein | Carbohydrate | Fat |
|---|---|---|
| (i) Increased catabolism<br>(ii) Increased utilization of BCAAs<br>(iii) Decreased ureagenesis | (i) Decreased hepatic and skeletal muscle glycogen synthesis<br>(ii) Increased gluconeogenesis<br>(iii) Glucose intolerance and insulin resistance | (i) Increased lipolysis<br>(ii) Enhanced turnover and oxidation of fatty acids<br>(iii) Increased Ketogenesis |

**Table 2:** Nutritional recommendations for the management of HE in end-stage liver failure

| Substrate | Recommendation |
|---|---|
| Energy | 35–40 kcal/kg/day |
| Protein | 1.2–1.5 g/kg of body weight/day* |
| BCAA | In severely protein-intolerant patients |
| Antioxidant and vitamins | Multivitamin supplements |
| Probiotics, prebiotics | Increasing use for ammonia-lowering and anti-inflammatory actions |

*In severely protein intolerant patients, protein may be reduced for short periods of time, particularly in grade III-IV hepatic encephalopathy.

It has been estimated that at least 25% of patients with liver cirrhosis experience HE during the natural history of the disease. HE is more frequent in patients with more severe liver insufficiency and in those with spontaneous or surgically created portal-systemic shunts. Whether or not malnourished patients are more prone to develop HE has not been clearly established, but could be anticipated based on several factors. Firstly, malnutrition tends to be more common in patients with advanced liver disease, and HE is more likely in this group. Secondly, nutritional deficits such as decreased lean body mass (muscle is important in ammonia uptake) and hypoalbuminemia (which increases free tryptophan levels) could promote HE [8].

# FACTORS CONTRIBUTING TO MALNUTRITION IN CIRRHOSIS

A range of factors are known to contribute to malnutrition in cirrhosis. These factors include (Figure 1) the following.

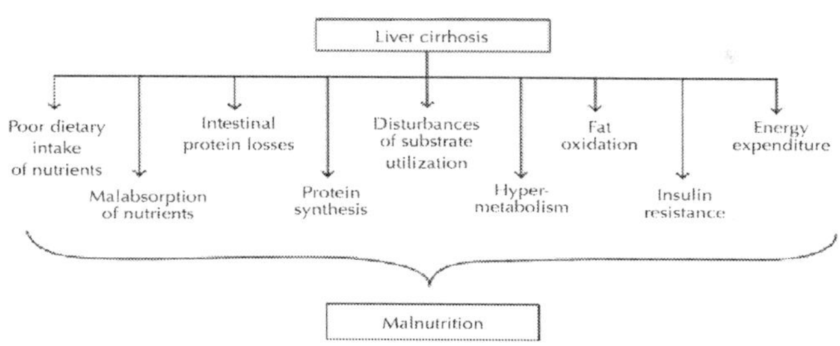


**Figure 1:** Factors contributing to malnutrition in end-stage liver failure.

## Inadequate Dietary Intake

Cirrhotic patients may unintentionally consume a low energy diet, an observation that is attributed to several factors including loss of appetite [9], anorexia, nausea, vomiting, early satiety, taste abnormalities, poor palatability of diets, reflux disease [10, 11], and impaired expansion capacity of the stomach [9].

## Inadequate Synthesis or Absorption of Nutrients

The cirrhotic liver may inadequately synthesize proteins and has diminished storage capacity and an impaired enterohepatic cycle. In addition, portal hypertensive enteropathy may lead to impaired absorption of essential nutrients. Moreover, pancreatic insufficiency, cholestasis, and drug-related diarrhea may all contribute to malabsorption in liver disease.

## Increased Protein Losses

Loss of proteins and minerals may result from complications of cirrhosis or from iatrogenic interventions such as the use of diuretics for the treatment of ascites and fluid retention as well as from the use of lactulose for the management of HE. Other potentially important causes of increased protein losses are blood loss from oesophageal and gastric varices and from the intestinal lumen due to ulcers or portal enteropathy.

## Hypermetabolic State/Increased Energy-Protein Expenditure and Requirements

The hyperdynamic circulation in cirrhosis leads to systemic vasodilation and an expanded intravascular blood volume. As a direct effect, a higher cardiac blood volume and therefore a greater use of macro- and micronutrients is a common cause of high energy expenditure and demand. Furthermore, the inability of the damaged liver to adequately clear activated proinflammatory mediators such as cytokines may promote the development of an inflammatory response with an increase in both energy expenditure and protein catabolism [12]. It has been suggested that elevated pro- and anti-inflammatory cytokine levels have the potential to result in hypermetabolism in cirrhosis [13, 14].

## Insulin Resistance

Insulin resistance and diabetes mellitus are common in patients with liver cirrhosis [15, 16]. Hyperinsulinemia and hyperglucagonemia are frequently present in cirrhotic patients where glucagon is disproportionately increased resulting in an elevated glucagon/insulin ratio. There is also impairment of glucose homeostasis due to hepatic insulin resistance characterized by altered gluconeogenesis, low glycogen stores, and impaired glycogenolysis [15, 16].

## Gastrointestinal Bleeding

Bleeding esophageal varices as a consequence of portal hypertension are frequent and severe complications of liver cirrhosis. Gastrointestinal bleeding is also a precipitating factor in HE and may accelerate progression of malnutrition in cirrhotic patients.

## Ascites

Impaired expansion capacity of the stomach due to the presence of clinically evident ascites may lead to an inadequate intake of nutrients [9], and cirrhotic patients with ascites often report early satiety and subsequent decreased oral intake which may result in significant weight loss [17].

## Inflammation/Infection

Malnourished patients with cirrhosis are prone to the development of inflammation and sepsis and their survival may be further shortened by these complications. There is a significant negative correlation between plasma levels of proinflammatory cytokines such as tumor necrosis factor-alpha (TNF-$\alpha$) and nutrient intake [18]. In order to reduce intestinal bacterial translocation and to improve gut immune function, it has been proposed that pre- and probiotics be added to the diet [19].

## Hyponatremia

Hyponatremia is a common complication of patients with advanced liver disease [20] and is an important predictor of short-term mortality. Hyponatremia is also an important pathogenic factor in patients with HE. Cirrhotic patients have abnormal sodium and water handling that may lead to refractory ascites. These patients retain sodium, and dilutional hyponatremia may develop, characterized by reduced serum sodium. In such situations, saline infusion should be avoided and it has been suggested that sodium intake should not exceed 2 g [21].

# ASSESSMENT OF NUTRITIONAL STATUS IN END-STAGE LIVER FAILURE

The nutritional assessment of the cirrhotic patient begins with the dietary history that should focus on nutritional intake and assessment of recent weight loss. However, altered mental status may preclude obtaining a meaningful history, and interviewing family members may be helpful.

Liver disease may interfere with biomarkers of malnutrition such as albumin, making it difficult to identify subjects at risk of malnutrition and to evaluate the need for nutritional intervention. Furthermore, anthropometric and bioelectrical impedance analysis may be biased by the presence of edema or ascites associated with liver failure. Body mass index (BMI), an index of nutritional status, may also be overvalued in patients with edema and ascites. Careful interpretation of nutritional data using these techniques in the presence of these complications is therefore required.

Generally accepted methods for assessing the clinical status and severity of disease in cirrhotic patients are the Child-Pugh-Turcotte classification [22] and the model for end-stage liver disease (MELD) [23, 24]. Unfortunately, these systems do not include an assessment of nutritional status in spite of the fact that malnutrition plays an important role in morbidity and mortality in end-stage liver failure. The omission of nutritional assessment results no doubt from the heterogeneous nature of the nutritional deficits in this population.

Subjective Global Assessment (SGA) and anthropometric parameters are the methods that are frequently used to evaluate nutritional status in end-stage liver failure [25]. SGA collects clinical information through history-taking, physical examination, and recent weight change and is considered to be reliable since it is minimally affected by fluid retention or the presence of ascites. The use of anthropometric parameters which are not affected by the presence of ascites or peripheral edema has also been recommended [22, 25]. Such parameters include mid-arm muscle circumference (MAMC), mid-arm circumference (MAC), and triceps skin fold thickness (TST). Diagnosis of malnutrition is established by values of MAMC and/or TST below the 5th percentile in patients aged 18–74 years, or the 10th percentile in patients aged over 74 years [26].

BMI changes may afford a reliable indicator of malnutrition using different BMI cutoff values depending on the presence and severity of ascites [26]; patients with a BMI below 22 with no ascites, below 23 with mild ascites, or below 25 with tense ascites are considered to be malnourished. Hand-grip examination by dynamometer has also been proposed as a simple method to detect patients at risk for the development of malnutrition [27]. In an interesting new development, Morgan et al. [28] validated a method where BMI and MAMC are combined with details of dietary intake in a semistructured algorithmic construct to provide a method for nutritional assessment in patients with end-stage liver failure [28]. Despite these advances, a standardized simple and accurate method for evaluating malnutrition in cirrhosis remains to be established.

## CONSEQUENCES OF CIRRHOSIS WITH A POTENTIAL TO IMPACT UPON NUTRITIONAL STATUS AND BRAIN FUNCTION

Cirrhosis results in multiple metabolic abnormalities and alterations in the synthesis, turnover, and elimination of a range of metal and micronutrients with the potential to alter nutritional status and consequently cerebral function. Such alterations include the following.

## Hyperammonemia

Under normal physiological conditions, ammonia is metabolized by the liver, brain, muscle, and kidney (Figure 2). In well-nourished cirrhotic patients, the affected liver has an impaired capacity for removal of ammonia in the form of urea, which may result in increased muscle glutamine synthetase in order to provide an alternative mechanism for ammonia removal as glutamine. Glutamine synthesis also increases to some extent in the brain of these patients. HE may develop as a consequence of increased circulating and cerebral ammonia in well-nourished cirrhotic patients. On the other hand, in malnourished cirrhotic patients, the loss of muscle mass, commonly seen as a consequence of malnutrition, can adversely affect this alternative route of ammonia removal. The brain being the main organ metabolizing ammonia in these conditions, severe HE is commonly diagnosed in malnourished cirrhotic patients.

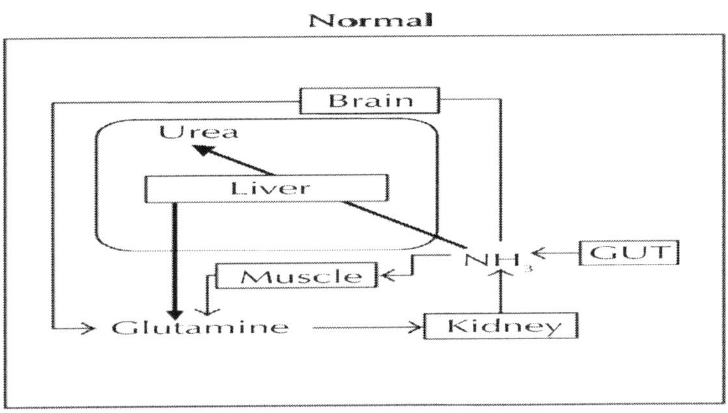


**Figure 2:** Inter-organ trafficking of ammonia in normal physiological conditions, in well-nourished patients with end-stage liver failure compared to malnourished end-stage liver failure patients.

Hyperammonemia may lead to increased uptake of tryptophan by the brain which may lead to increased synthesis and release of serotonin and anorexia. This symptom may render the patient prone to chronic catabolism and malnutrition, and in turn to increased ammonia load, resulting in a vicious cycle [29, 30]. In addition, hyperammonemia may be more prominent after gastrointestinal bleeding due to the absence of isoleucine [31]. Since haemoglobin molecule lacks the essential amino acid isoleucine, gastrointestinal bleed may stimulate the induction of net catabolism [32].

## Zinc

Zinc is an essential trace element that plays an important role in the regulation of protein and nitrogen metabolism as well as in antioxidant defense. Reduced zinc content is common in cirrhotic patients, but zinc deficiency cannot be effectively diagnosed based upon serum concentrations since zinc is bound to albumin, which is also decreased in these patients [33, 34]. Among the mechanisms contributing to zinc deficiency, poor dietary intake [35], reduced intestinal absorption [36], reduced hepatointestinal extraction [37], portal-systemic shunting, and altered protein and amino acid metabolism have all been implicated [38]. Zinc deficiency may impair the activity of enzymes of the urea cycle as well as glutamine synthetase [39, 40], and decreased activity of these enzymes has the potential to lead to further increases in circulating and brain ammonia with the potential to cause worsening of HE. Not surprisingly, therefore, an inverse relationship between serum zinc and ammonia concentrations has been described [41, 42]. Zinc deficiency has been implicated in multiple complications of cirrhosis, including poor appetite, immune dysfunction, altered taste and smell, anorexia as well as altered protein metabolism [43, 44]. Surprisingly, in spite of evidence of hypozincemia in cirrhosis, zinc supplementation in the treatment of HE based on a small number of controlled trials has so far provided inconsistent results, a finding that may be attributable to variations in the nature and doses of zinc salts used and to duration of therapy [45].

## Selenium

Decreased levels of selenium have been reported in cirrhotic patients [46, 47]. However, the relationship of diminished selenium to the pathogenesis of cirrhosis and its complications, including HE, has not been clearly established.

## Manganese

In cirrhotic patients, the elimination of manganese is decreased secondary to impaired hepatobiliary function and portal-systemic shunting, which result in increased blood manganese levels and increased manganese deposition in basal ganglia structures of the brain, in particular in globus pallidus [48–52]. Manganese has also been correlated to increased brain glutamine levels [53] and changes in dopamine metabolism [49, 54] and may be related to other alterations in cirrhotic patients with HE, such as the characteristic astrocytic morphologic changes [55]. Toxic effects of manganese on central nervous system could be mediated by effects on the glycolytic enzyme glyceraldehyde-3-phosphate dehydrogenase (GAPDH) [56]. It was also suggested that manganese-induced increases of "peripheral-type" benzodiazepine receptors (PTBRs) could contribute to the pathogenesis of HE [57].

## L-Carnitine

The liver is a major site for the production of ketone bodies from the oxidation of fatty acids. Fatty acids cannot penetrate the inner mitochondrial matrix and cross the mitochondrial membrane to undergo oxidation unless they are transported by a carrier process involving L-carnitine (3-hydroxy-4-trimethylammoniobutanoate). Carnitine is a cofactor for mitochondrial oxidation of fatty acids and prevents the body from using fats for energy production particularly during starvation. Carnitine deficiency may result in lethargy, somnolence, confusion, and encephalopathy. Studies of carnitine status in cirrhotic patients have yielded conflicting results; the source of this lack of consensus likely results from both the etiology of cirrhosis and the severity of liver disease. For example, Rudman et al. [58] reported reduced plasma and tissue carnitine concentrations in patients with alcoholic cirrhosis complicated by cachexia, whereas later studies by Fuller and Hoppel [59, 60] reported an increase of plasma carnitine in alcoholics with or without cirrhosis. De Sousa et al. [61] reported no such changes in a similar patient population. In a subsequent study by Amodio et al. [62], plasma carnitine levels were measured in cirrhotic patients and the relationship to nutritional status and severity of liver damage was assessed. Plasma carnitine levels did not differ between Child-Pugh class A, B, and C patients. Significantly higher levels of acetylcarnitine, short chain acylcarnitine, total esterified carnitine, and total carnitine were observed in cirrhotic patients independent of etiology of cirrhosis. The issue of carnitine in relation to liver disease was re-evaluated in 1997 by Krähenbühl and Reichen [63] who studied carnitine metabolism in 29 patients with chronic liver disease of varying degrees of severity and various etiologies. Patients with alcoholic cirrhosis manifested increased total plasma carnitine levels with a close correlation to serum bilirubin. Urinary carnitine excretion was not different between cirrhotic patients and controls with the exception of patients with primary biliary cirrhosis. It was concluded that patients with cirrhosis are not normally carnitine deficient and that patients with alcohol-induced cirrhosis manifest hypercarnitinemia which results primarily from increased carnitine synthesis due to increased skeletal muscle protein turnover [63].

## Vitamin $B_1$ (Thiamine)

Wernicke's Encephalopathy caused by vitamin $B_1$ deficiency and characterized by a triad of neurological symptoms (ophthalmoplegia, ataxia, global confusional state) is common in cirrhotic patients. In a retrospective neuropathological study of sections from patients with end-stage liver failure who died in hepatic coma, 64% were found to manifest thalamic lesions typical of Wernicke's Encephalopathy [64]. None of the cases of Wernicke's Encephalopathy had been suspected based upon clinical symptoms during life, a finding which draws into question the classical textbook definition based upon symptomatology associated with the disorder [65].

Causes of vitamin $B_1$ deficiency in cirrhosis include reduced dietary intake, impaired absorption, and loss of hepatic stores of the vitamin. Alcoholic cirrhotic patients manifest increased incidence of vitamin $B_1$deficiency compared to nonalcoholic cirrhotics [66]. Moreover, ethanol is known to impair both intestinal absorption of vitamin $B_1$ [67] and to impair the transformation of the vitamin into its active (diphosphorylated) form [68]. It has been suggested that common pathophysiologic mechanisms exist in Wernicke's and hepatic encephalopathies, related to deficits of vitamin $B_1$-dependent enzymes [69]. Vitamin $B_1$ supplementation is highly recommended in patients with end-stage liver failure of either alcoholic or nonalcoholic etiologies.

## NUTRITION, HE, AND LIVER TRANSPLANTATION

HE in end-stage liver failure may contribute to malnutrition in the pretransplant period as a consequence of diminished food uptake [70]. Alterations in markers of nutritional status such as serum albumin are significant risk factors for both surgical [71] and postsurgical [72] complications of liver transplantation. Moreover, it has been suggested that nonabsorbable disaccharides (such as lactulose) administered for the management of HE may result in intestinal malabsorption in patients with end-stage liver failure with the potential to result in poor transplant outcome [73].

The negative impact of malnutrition on liver transplantation had been reported in early retrospective studies [74]. Both preoperative hypermetabolism and body cell mass depletion proved to be of prognostic value for transplantation outcome [75]. Malnutrition is known to lead to glycogen depletion, and this has been suggested to increase the plasma lactate:pyruvate ratio during the anhepatic phase and to induce an exacerbated proinflammatory cytokine response, thereby favouring the development of postoperative systemic inflammatory response syndrome and multiorgan failure in these patients [76]. To date, there are still insufficient data in the pretransplant period upon which to base specific recommendations. In the posttransplant period, nutritional therapy improves nitrogen balance, decreases viral infection, and shows a trend to shortened intensive care unit stays with lowering of hospitalisation costs [77, 78].

# NUTRITIONAL RECOMMENDATIONS FOR HE IN END-STAGE LIVER FAILURE (TABLE 2)

## General Considerations

Considering the high prevalence of malnutrition in cirrhotic patients together with the lack of simple and accurate methods of assessment of malnutrition in this patient population, it is reasonable to assume that malnutrition occurs in all patients. Nutritional requirements may vary according to the specific clinical situation. Multiple (5-6) small feedings with a carbohydrate-rich evening snack have been recommended with complex rather than simple carbohydrates used for calories. Lipids could provide 20%–40% of caloric needs. Long-term nutritional supplements may be necessary to provide recommended caloric and protein requirements. Additional studies are needed in order to formulate specific recommendations for nutrients such as zinc, selenium, and carnitine.

## Energy Requirements

The primary goal for a patient suffering from end-stage liver failure should be to avoid by all means possible intentional or unintentional weight loss and sustain a diet rich in nutrients. It has been suggested that patients with liver cirrhosis should receive 35–40 kcal/kg per day [25].

## Low Protein Diet to be Avoided

Restriction of dietary protein was long considered a mainstay in the management of liver disease and HE [79,80]. In particular, protein restriction (0–40 g protein/day) was shown to decrease encephalopathy grade in patients following surgical creation of a portal-systemic shunt, the only available therapy at one time for bleeding varices. Protein restriction (0–40 g protein/day) was later extended to include all patients with cirrhosis who developed encephalopathy. However, more recently, studies have shown that protein restriction in these patients has no impact on encephalopathy grade and that it may even worsen their nutritional status [81]. The increased awareness of the progressive deterioration of nutritional status in liver cirrhosis combined with a better understanding of metabolic alterations in the disorder has questioned the practice of prolonged protein restriction in the management of HE [82]. In fact, protein requirements are increased in cirrhotic patients, and high protein diets are generally well tolerated in the majority of patients. Moreover, the inclusion of adequate protein in the diets of malnourished patients with end-stage liver failure is often associated with a sustained improvement in their mental status. Furthermore, protein helps preserve lean body mass; this is crucial in patients with liver failure in whom skeletal muscle makes a significant contribution to ammonia removal. The consensus of opinion nowadays is that protein restriction be avoided in all but a small number of patients with severe protein intolerance and that protein be maintained between 1.2 and 1.5 g of proteins per kg of body weight per day. In severely protein intolerant patients, particularly in patients in grades III-IV HE, protein may be reduced for short periods of time [83–85].

## Vegetable Versus Animal Proteins

It has been suggested that vegetable proteins are better tolerated than animal proteins in patients with end-stage liver failure, a finding that has been attributed to either their higher content of branched-chain amino acids and/or because of their influence on intestinal transit [86, 87]. One study reported that a diet rich in vegetable protein (71 g/d) significantly improved the mental status of patients suffering from HE while increasing their nitrogen balance [88]. Vegetable proteins may also increase intraluminal pH and decrease gastric transit time. High dietary fibre diet has been recommended in order to abolish constipation which is an established precipitating factor for HE in patients with cirrhosis [89, 90]. A daily intake of 30–40 g vegetable protein has been found to be effective in the majority of patients [88].

## Branched-Chain Amino Acids (Bcaas)

These amino acids (leucine, isoleucine, and valine) cannot be synthesized de novo but must be obtained from dietary sources and have a unique role in amino acid metabolism, regulating the intra- and interorgan exchange of nitrogen and amino acids by different tissues [91]. Chronic liver disease and portal-systemic shunting are characterized by a decrease in the plasma concentrations of BCAAs [92], whereas hyperammonemia increases their utilization. Since hyperammonemia results in increased utilization of BCAAs, which are largely metabolized by the muscle, it would be anticipated that providing BCAAs could facilitate ammonia detoxification by supporting muscle glutamine synthesis. Administration of BCAAs has been shown to stimulate hepatic protein synthesis; indeed, leucine stimulates the synthesis of hepatocyte growth factor by stellate cells [93]. Also, BCAAs reduce postinjury catabolism and improve nutritional status. Inadequate dietary protein intake or low levels of BCAAs may have a deleterious effect on HE [94], nutritional status [80], and clinical outcome [25, 81] in patients with end-stage liver failure. Clinical trials of BCAAs in the treatment of HE have yielded inconsistent findings. Several controlled clinical studies reported no efficacy of BCAAs on encephalopathy grade in patients with cirrhosis [95, 96]. However, other trials demonstrated that BCAAs were beneficial in similar patients [97, 98].

A double-blind, randomized clinical trial demonstrated that, in advanced cirrhosis, long-term nutritional supplementation with oral BCAA was useful to prevent progressive hepatic failure [99]. Furthermore, administration of solutions enriched with BCAAs has been shown to improve cerebral perfusion in cirrhotic patients [100]. Muto et al. [101] confirmed the beneficial effects of BCAAs using a more palatable granular formula. In a multicenter randomized study, it was also reported that long-term oral supplementation with a BCAA mixture improved the serum albumin level as well as cellular energy metabolism in cirrhotic patients [102].

The timing of BCAA supplementation in patients with end-stage liver failure may be crucial. This issue was addressed by a crossover study of 12 cirrhotic patients [103]. Daytime administration improved nitrogen balance and Fischer's ratio (ratio of BCAA/AAAs); however, both were further improved with nocturnal administration. At 3 months, a significant increase in serum albumin level was observed in patients administered nocturnal BCAAs, but not daytime BCAAs. It is possible that daytime BCAAs may be used primarily as calories, whereas nocturnal BCAAs may be preferentially used for protein synthesis. Furthermore, the long-term use of BCAAs in liver cirrhosis leads to an increase of serum protein of approximately 10% if given before bedtime [104]. Problems that limit the widespread use of BCAAs in the treatment of HE include their expense and unpalatability [105], both of which may result in poor patient compliance.

# ANTIOXIDANTS

## Rationale for Use of Antioxidants

Cirrhotic patients manifest evidence of increased expression of biomarkers of oxidative stress such as increased lipid peroxidation [106, 107], as well as impaired antioxidant defences. Decreased levels of antioxidant micronutrients, including zinc [33, 107], selenium [46, 47], and vitamin E [107, 108] have been described in patients with end-stage liver failure. The potential benefits of vitamin E have been investigated, but results are conflicting. One randomized, placebo-controlled trial of vitamin E supplementation revealed a significant amelioration in terms of liver inflammation and fibrosis in patients with nonalcoholic steatohepatitis [109], while other studies with biochemical end points did not demonstrate any significant beneficial effect of vitamin E supplements [110]. In an earlier placebo-controlled randomized trial, 1-year vitamin E supplementation to patients with end-stage liver failure led to increased serum alpha-tocopherol levels, but did not result in any improvement in survival or quality of life [111]. The benefits of vitamin E therapy in relation to HE have not been assessed.

## N-Acetylcysteine

A widely used complementary medical therapy for acute liver failure is the glutathione prodrug, N-acetylcysteine (NAC) [112, 113].

Glutathione is a major component of the pathways by which cells are protected from oxidative stress. NAC is an antioxidant with a thiol-containing compound and is used to restore cytosolic glutathione and detoxify reactive oxygen species and free radicals. NAC has proven beneficial in patients with type I hepatorenal syndrome [112] but was inefficient in patients with hepatitis C [113]. While NAC is widely used to treat acetaminophen hepatotoxicity, its benefit in end-stage liver failure with specific reference to HE remains to be established. In this regard, NAC is known to cross the blood-brain barrier and to improve central antioxidant status in the brain in mice with acute liver failure due to azoxymethane-induced hepatotoxicity [114].

# WATER-SOLUBLE AND FAT-SOLUBLE VITAMINS

Deficiencies in water-soluble vitamins (particularly the vitamin B complex) are common in end-stage liver failure [115]. A wide range of neuropsychiatric symptomatology associated with liver disease may be the consequence of water-soluble vitamin deficiencies. For example, peripheral neuropathy may result from pyridoxine, thiamine, or vitamin $B_{12}$ deficiency. Confusion, ataxia and ocular disturbances are cardinal features of a lack of thiamine, and thiamine deficiency has been reported in patients with hepatitis C-related cirrhosis [116]. Deficiencies in vitamin $B_{12}$, thiamine, and folic acid may develop faster in cirrhotic patients due to diminished hepatic storage.

Fat-soluble vitamins (A, D, and K) deficiencies are likely to arise from malabsorption associated with end-stage liver failure. Vitamin A supplementation may be considered since vitamin A deficiency results in nyctalopia and dry cornea, and is associated with increased risk of hepatocellular carcinoma in patients with end-stage liver disease [117, 118]. Prescription of vitamin D, especially in patients with cholestasis (in combination with calcium since osteoporosis may be a complication of end-stage liver failure), is advised [118, 119]. Also, supplementation of vitamin K in conditions with high risk of bleeding such as the presence of impaired prothrombin time and oesophageal varices, should be considered [118]. In view of these findings, administration of multivitamin preparations is recommended.

# PROBIOTICS, PREBIOTICS, AND SYNBIOTICS

Probiotics are live microbiological dietary supplements with beneficial effects on the host beyond their nutritional properties. Prebiotics stimulate the growth and activity of beneficial bacteria within the intestinal flora. Synbiotics are a combination of pro- and prebiotics. Their mechanisms of action include the deprivation of substrates for potentially pathogenic bacteria, together with the provision of fermentation end products for potentially beneficial bacteria. Probiotic or prebiotic treatments aim at increasing the intestinal content of lactic acid-type bacteria at the expense of other species with more pathogenic potential.

The concept of treating HE with probiotics was already suggested several decades ago [120–122]. The therapeutic benefit of acidifying the gut lumen with synbiotics in cirrhotic patients with minimal HE was demonstrated by Liu et al. [123] who showed that synbiotic/probiotic supplementation ameliorates hepatic function as reflected by reduced bilirubin and albumin levels and prothrombin times [123]. Modulation of gut flora was also associated with a significant reduction in blood ammonia levels and a reversal of minimal HE in 50% of patients [123]; improved hepatic function and serum transaminase levels in patients with alcohol- and hepatitis C-related cirrhosis have also been reported [124]. Another group reported improvement in biochemical and neuropsychological tests in cirrhotic patients receiving probiotics [125, 126]. Furthermore, liver transplant recipients who received a synbiotic regimen developed significantly fewer bacterial infections [127]. In a subsequent clinical trial, the incidence of postoperative bacterial infection as well as the duration of antibiotic therapy was significantly reduced in liver transplant patients receiving prebiotics [128]. More recently, Bajaj et al. [129] demonstrated a significant rate of minimal HE reversal in cirrhotic patients after probiotic yogurt supplements. Probiotics may provide additional benefits over dietary supplementation in reducing episodes of infection. Given the efficacy of probiotics and their lack of side effects, they are increasingly being used in the management of HE..

# CONCLUSIONS

Malnutrition is common in patients with end-stage liver failure and HE and adversely affects prognosis. Inadequate dietary intake, altered synthesis and absorption of nutrients, increased protein losses, hypermetabolism, and inflammation are among the factors contributing to malnutrition in this patient population. Although there are now several available methods to assess malnutrition, a standardized simple and accurate method for evaluating malnutrition in end-stage liver failure remains a challenge. Consequences of end-stage liver failure with a potential to impact upon nutritional status and brain function are numerous and include hyperammonemia, reduced zinc and selenium, manganese accumulation as well as deficiencies of carnitine and water-soluble vitamins, particularly thiamine. The primary goal for a patient with end-stage liver failure is to avoid by all means possible weight loss and sustain a diet rich in nutrients. A caloric intake of 35–40 kcal/kg/day is recommended. Low protein diets should be avoided and protein intake maintained at 1.2–1.5 g/kg/day. Particular attention should also be drawn to vegetable protein as well as to BCAAs which have proven beneficial in the treatment of HE. Antioxidants as well as probiotics are increasingly being employed in order to optimize the nutritional status in cirrhotic patients. Administration of multivitamin preparations, particularly thiamine, is recommended for patients with end-stage liver failure. Nutritional support to meet energy and substrate needs and to optimize the removal of circulating ammonia, reduce proinflammatory mechanisms, and improve antioxidant defenses has the potential to limit the progression of liver dysfunction, treat HE, and improve quality of life in patients with end-stage liver failure.

# REFERENCES

1. Italian Multicentre Cooperative Project on Nutrition in Liver Cirrhosis, "Nutritional status in cirrhosis," Journal of Hepatology, vol. 21, no. 3, pp. 317–325, 1994.
2. M. J. Müller, "Malnutrition in cirrhosis," Journal of Hepatology, Supplement, vol. 23, no. 1, pp. 31–35, 1995.
3. L. Caregaro, F. Alberino, P. Amodio, et al., "Malnutrition in

alcoholic and virus-related cirrhosis,"American Journal of Clinical Nutrition, vol. 63, no. 4, pp. 602–609, 1996.

4. C. Mendenhall, G. A. Roselle, P. Gartside, and T. Moritz, "Relationship of protein calorie malnutrition to alcoholic liver disease: a reexamination of data from two Veteran Administration Cooperative Studies," Alcoholism: Clinical and Experimental Research, vol. 19, no. 3, pp. 635–641, 1995.
5. B. Campillo, J. P. Richardet, E. Scherman, and P. N. Bories, "Evaluation of nutritional practice in hospitalized cirrhotic patients: results of a prospective study," Nutrition, vol. 19, no. 6, pp. 515–521, 2003.
6. K. K. Changani, R. Jalan, I. J. Cox et al., "Evidence for altered hepatic gluconeogenesis in patients with cirrhosis using in vivo 31-phosphorus magnetic resonance spectroscopy," Gut, vol. 49, no. 4, pp. 557–564, 2001.
7. B. Campillo, C. Chapelain, J. C. Bonnet et al., "Hormonal and metabolic changes during exercise in cirrhotic patients," Metabolism, vol. 39, no. 1, pp. 18–24, 1990.
8. T. L. Sourkes, "Tryptophan in hepatic coma," Journal of Neural Transmission, Supplement, no. 14, pp. 79–86, 1978.
9. M. Plauth and E. T. Schütz, "Cachexia in liver cirrhosis," International Journal of Cardiology, vol. 85, no. 1, pp. 83–87, 2002.
10. F. Izbéki, I. Kiss, T. Wittmann, T. T. Várkonyi, P. Légrády, and J. Lonovics, "Impaired accommodation of proximal stomach in patients with alcoholic liver cirrhosis," Scandinavian Journal of Gastroenterology, vol. 37, no. 12, pp. 1403–1410, 2002.
11. B. A. Aqel, J. S. Scolapio, R. C. Dickson, D. D. Burton, and E. P. Bouras, "Contribution of ascites to impaired gastric function and nutritional intake in patients with cirrhosis and ascites," Clinical Gastroenterology and Hepatology, vol. 3, no. 11, pp. 1095–1100, 2005.
12. M. A. Boermeester, A. P. Houdijk, S. Meyer et al., "Liver failure induces a systemic inflammatory response: prevention by recombinant N-terminal bactericidal/permeability-increasing protein,"American Journal of Pathology, vol. 147, no. 5, pp. 1428–1440, 1995.

13. H. Tilg, A. Wilmer, W. Vogel et al., "Serum levels of cytokines in chronic liver diseases,"Gastroenterology, vol. 103, no. 1, pp. 264–274, 1992.

14. V. von Baehr, W. D. Döcke, M. Plauth et al., "Mechanisms of endotoxin tolerance in patients with alcoholic liver cirrhosis: role of interleukin 10, interleukin 1 receptor antagonist, and soluble tumour necrosis factor receptors as well as effector cell desensitisation," Gut, vol. 47, no. 2, pp. 281–287, 2000.·

15. A. J. McCullough and A. S. Tavill, "Disordered energy and protein metabolism in liver disease,"Seminars in Liver Disease, vol. 11, no. 4, pp. 265–277, 1991.

16. T. Nishida, S. Tsuji, M. Tsujii et al., "Oral glucose tolerance test predicts prognosis of patients with liver cirrhosis," American Journal of Gastroenterology, vol. 101, no. 1, pp. 70–75, 2006.

17. J. S. Scolapio, A. Ukleja, K. McGreevy, O. L. Burnett, and P. C. O'Brien, "Nutritional problems in end-stage liver disease: contribution of impaired gastric emptying and ascites," Journal of Clinical Gastroenterology, vol. 34, no. 1, pp. 89–93, 2002.

18. C. Gerstner, T. Schuetz, V. von Baehr, et al., "Correlation between energy expenditure, nutrient intake, malnutrition and activation of the inflammatory system in patients with liver cirrhosis," Journal of Hepatology, vol. 34, supplement 1, pp. 195–196, 2001.

19. G. Bianchi, R. Marzocchi, C. Lorusso, V. Ridolfi, and G. Marchesini, "Nutritional treatment of chronic liver failure," Hepatology Research, vol. 38, no. 1, pp. S93–S101, 2008.

20. E. Bartoli, L. Castello, and P. P. Sainaghi, "Diagnosis and therapy of hyponatremia," Annali Italiani di Medicina Interna, vol. 18, no. 4, pp. 193–203, 2003.

21. N. Tözün, "Influence of the metabolic complications of liver cirrhosis on dietary intake," Medical Science Monitor, vol. 6, no. 6, pp. 1223–1226, 2000.

22. F. Gunsar, M. L. Raimondo, S. Jones et al., "Nutritional status and prognosis in cirrhotic patients,"Alimentary Pharmacology and Therapeutics, vol. 24, no. 4, pp. 563–572, 2006.

23. L. M. Forman and M. R. Lucey, "Predicting the prognosis of chronic liver disease: an evolution from child to MELD," Hepatology, vol. 33, no. 2, pp. 473–475, 2001.

24. L. Pagliaro, "MELD: the end of Child-Pugh classification?" Journal of Hepatology, vol. 36, no. 1, pp. 141–142, 2002.
25. M. Plauth, E. Cabré, O. Riggio et al., "ESPEN Guidelines on Enteral Nutrition: liver disease," Clinical Nutrition, vol. 25, no. 2, pp. 285–294, 2006.
26. B. Campillo, J. P. Richardet, and P. N. Bories, "Enteral nutrition in severely malnourished and anorectic cirrhotic patients in clinical practice: benefit and prognostic factors," Gastroenterologie Clinique et Biologique, vol. 29, no. 6-7, pp. 645–651, 2005.
27. M. R. Álvares-da-Silva and T. Reverbel da Silveira, "Comparison between handgrip strength, subjective global assessment, and prognostic nutritional index in assessing malnutrition and predicting clinical outcome in cirrhotic outpatients," Nutrition, vol. 21, no. 2, pp. 113–117, 2005.
28. M. Y. Morgan, A. M. Madden, C. T. Soulsby, and R. W. Morris, "Derivation and validation of a new global method for assessing nutritional status in patients with cirrhosis," Hepatology, vol. 44, no. 4, pp. 823–835, 2006.
29. C. Bachmann, O. Braissant, A. M. Villard, O. Boulat, and H. Henry, "Ammonia toxicity to the brain and creatine," Molecular Genetics and Metabolism, vol. 81, pp. S52–S57, 2004.
30. R. M. Cohn and K. S. Roth, "Hyperammonemia, bane of the brain," Clinical Pediatrics, vol. 43, no. 8, pp. 683–689, 2004.
31. C. H. Dejong, W. J. Meijerink, C. L. van Berlo, N. E. Deutz, and P. B. Soeters, "Decreased plasma isoleucine concentrations after upper gastrointestinal haemorrhage in humans," Gut, vol. 39, no. 1, pp. 13–17, 1996.
32. S. W. Olde Damink, C. H. Dejong, and R. Jalan, "Review article: hyperammonaemic and catabolic consequences of upper gastrointestinal bleeding in cirrhosis," Alimentary Pharmacology and Therapeutics, vol. 29, no. 8, pp. 801–810, 2009.
33. K. Kisters, C. Spieker, S. Q. Nguyen, H. P. Bertram, C. Muller, and W. Zidek, "Zinc concentrations in human liver tissue and in blood plasma in cirrhosis of the liver due to alcoholism," Trace Elements and Electrocytes, vol. 11, no. 3, pp. 101–103, 1994.
34. H. F. Goode, J. Kelleher, and B. E. Walker, "Relation between zinc status and hepatic functional reserve in patients with liver

disease," Gut, vol. 31, no. 6, pp. 694–697, 1990.

35. L. Capocaccia, M. Merli, C. Piat, R. Servi, A. Zullo, and O. Riggio, "Zinc and other trace elements in liver cirrhosis," Italian Journal of Gastroenterology, vol. 23, no. 6, pp. 386–391, 1991.

36. K. Gohshi, S. Fujiyama, J. Shibata, T. Sato, A. Higashi, and I. Matsuda, "Zinc absorption and its correlation with results of oral zinc tolerance testing in non-alcoholic liver cirrhosis; kinetic study,"Hepatogastroenterology, vol. 42, no. 5, pp. 487–491, 1995.

37. P. W. Keeling, W. Ruse, J. Bull, B Hannigan, and R. P. Thompson, "Direct measurement of the hepatointestinal extraction of zinc in cirrhosis and hepatitis," Clinical Science, vol. 61, no. 4, pp. 441–444, 1981.

38. E. Rocchi, P. Borella, A. Borghi et al., "Zinc and magnesium in liver cirrhosis," European Journal of Clinical Investigation, vol. 24, no. 3, pp. 149–155, 1994.

39. O. Riggio, M. Merli, L. Capocaccia et al., "Zinc supplementation reduces blood ammonia and increases liver ornithine transcarbamylase activity in experimental cirrhosis," Hepatology, vol. 16, no. 3, pp. 785–789, 1992.

40. A. S. Prasad, P. Rabbani, A. Abbasii, E. Bowersox, and M. R. Fox, "Experimental zinc deficiency in humans," Annals of Internal Medicine, vol. 89, no. 4, pp. 483–490, 1978.

41. P. Reding, J. Duchateau, and C. Bataille, "Oral zinc supplementation improves hepatic encephalopathy. Results of a randomised controlled trial," Lancet, vol. 2, no. 8401, pp. 493–495, 1984.

42. K. Grüngreiff, H. J. Presser, D. Franke, B. Lössner, K. Abicht, and F. D. Kleine, "Correlations between zinc, amino acids and ammonia in liver cirrhosis," Zeitschrift fur Gastroenterologie, vol. 27, no. 12, pp. 731–735, 1989.

43. A. S. Prasad, "Effects of zinc deficiency on Th1 and Th2 cytokine shifts," Journal of Infectious Diseases, vol. 182, no. 3, pp. S62–S68, 2000.

44. G. Marchesini, A. Fabbri, G. Bianchi, M. Brizi, and M. Zoli, "Zinc supplementation and amino acid-nitrogen metabolism in patients with advanced cirrhosis," Hepatology, vol. 23, no. 5, pp.

1084–1092, 1996.

45. P. Marchetti, P. Amodio, L. Caregaro, and A. Gatta, "Zinc deficiency in liver cirrhosis: a curiosity or problem?" Annali Italiani di Medicina Interna, vol. 13, no. 3, pp. 157–162, 1998.
46. B. Dworkin, W. S. Rosenthal, R. H. Jankowski, G. G. Gordon, and D. Haldea, "Low blood selenium levels in alcoholics with and without advanced liver disease. Correlations with clinical and nutritional status," Digestive Diseases and Sciences, vol. 30, no. 9, pp. 838–844, 1985.
47. A. Van Gossum and J. Nève, "Low selenium status in alcoholic cirrhosis is correlated with aminopyrine breath test: preliminary effects of selenium supplementation," Biological Trace Element Research, vol. 47, no. 1–3, pp. 201–207, 1995.
48. G. Pomier-Layrargues, L. Spahr, and R. F. Butterworth, "Increased manganese concentrations in pallidum of cirrhotic patients," Lancet, vol. 345, no. 8951, p. 735, 1995.
49. R. F. Butterworth, L. Spahr, S. Fontaine, and G. P. Layrargues, "Manganese toxicity, dopaminergic dysfunction and hepatic encephalopathy," Metabolic Brain Disease, vol. 10, no. 4, pp. 259–267, 1995.·
50. L. Spahr, R. F. Butterworth, S. Fontaine et al., "Increased blood manganese in cirrhotic patients: relationship to pallidal magnetic resonance signal hyperintensity and neurological symptoms,"Hepatology, vol. 24, no. 5, pp. 1116–1120, 1996.
51. C. Rose, R. F. Butterworth, J. Zayed et al., "Manganese deposition in basal ganglia structures results from both portal-systemic shunting and liver dysfunction," Gastroenterology, vol. 117, no. 3, pp. 640–644, 1999.
52. G. P. Layrargues, D. Shapcott, L. Spahr, and R. F. Butterworth, "Accumulation of manganese and copper in pallidum of cirrhotic patients: role in the pathogenesis of hepatic encephalopathy?" Metabolic Brain Disease, vol. 10, no. 4, pp. 353–356, 1995.
53. S. Montes, M. Alcaraz-Zubeldia, P. Muriel, and C. Rios, "Role of manganese accumulation in increased brain glutamine of the cirrhotic rat," Neurochemical Research, vol. 28, no. 6, pp. 911–917, 2003.
54. S. C. Sistrunk, M. K. Ross, and N. M. Filipov, "Direct effects of manganese compounds on dopamine and its metabolite Dopac:

an in vitro study," Environmental Toxicology and Pharmacology, vol. 23, no. 3, pp. 286–296, 2007.

55. R. F. Butterworth, "Complications of cirrhosis. III. Hepatic encephalopathy," Journal of Hepatology, vol. 32, no. 1, pp. 171–180, 2000.
56. A. S. Hazell, P. Desjardins, and R. F. Butterworth, "Increased expression of glyceraldehyde-3-phosphate dehydrogenase in cultured astrocytes following exposure to manganese," Neurochemistry International, vol. 35, no. 1, pp. 11–17, 1999.
57. A. S. Hazell, P. Desjardins, and R. F. Butterworth, "Chronic exposure of rat primary astrocyte cultures to manganese results in increased binding sites for the 'peripheral-type' benzodiazepine receptor ligandH3-PK 11195," Neuroscience Letters, vol. 271, no. 1, pp. 5–8, 1999.
58. D. Rudman, C. W. Sewell, and J. D. Ansley, "Deficiency of carnitine in cachectic cirrhotic patients,"Journal of Clinical Investigation, vol. 60, no. 3, pp. 716–723, 1977.
59. R. K. Fuller and C. L. Hoppel, "Elevated plasma carnitine in hepatic cirrhosis," Hepatology, vol. 3, no. 4, pp. 554–558, 1983.
60. R. K. Fuller and C. L. Hoppel, "Plasma carnitine in alcoholism," Alcoholism: Clinical and Experimental Research, vol. 12, no. 5, pp. 639–642, 1988.
61. C. De Sousa, N. W. Leung, R. A. Chalmers, andT. J. Peters, "Free and total carnitine and acylcarnitine content of plasma, urine, liver and muscle of alcoholics," Clinical Science, vol. 75, no. 4, pp. 437–440, 1988.
62. P. Amodio, P. Angeli, C. Merkel, F. Menon, and A. Gatta, "Plasma carnitine levels in liver cirrhosis: relationship with nutritional status and liver damage," Journal of Clinical Chemistry and Clinical Biochemistry, vol. 28, no. 9, pp. 619–626, 1990.
63. S. Krähenbühl and J. Reichen, "Carnitine metabolism in patients with chronic liver disease,"Hepatology, vol. 25, no. 1, pp. 148–153, 1997.
64. J. J. Kril and R. F. Butterworth, "Diencephalic and cerebellar pathology in alcoholic and nonalcoholic patients with end-stage liver disease," Hepatology, vol. 26, no. 4, pp. 837–841, 1997.

65. C. G. Harper and R. F. Butterworth, "Nutritional and metabolic disorders," in Greenfield›s Neuropathology, pp. 601–655, Hodder Arnold, 6th edition, 1997.
66. U. Laforenza, C. Patrini, G. Gastaldi, and G. Rindi, "Effects of acute and chronic ethanol administration on thiamine metabolizing enzymes in some brain areas and in other organs of the rat," Alcohol and Alcoholism, vol. 25, no. 6, pp. 591–603, 1990.
67. A. D. Thomson, H. Baker, and C. M. Leevy, "Patterns of S-thiamine hydrochloride absorption in the malnourished alcoholic patient," The Journal of Laboratory and Clinical Medicine, vol. 76, no. 1, pp. 34–45, 1970.
68. G. Rindi, L. Imarisio, and C. Patrini, "Effects of acute and chronic ethanol administration on regional thiamin pyrophosphokinase activity of the rat brain," Biochemical Pharmacology, vol. 35, no. 22, pp. 3903–3908, 1986.
69. R. F. Butterworth, "Thiamine deficiency-related brain dysfunction in chronic liver failure," Metabolic Brain Disease, vol. 24, no. 1, pp. 189–196, 2009.
70. A. S. Henkel and A. L. Buchman, "Nutritional support in patients with chronic liver disease," Nature Clinical Practice Gastroenterology and Hepatology, vol. 3, no. 4, pp. 202–209, 2006.
71. S. A. McCluskey, K. Karkouti, D. N. Wijeysundera et al., "Derivation of a risk index for the prediction of massive blood transfusion in liver transplantation," Liver Transplantation, vol. 12, no. 11, pp. 1584–1593, 2006.
72. G. Fusai, P. Dhaliwal, N. Rolando et al., "Incidence and risk factors for the development of prolonged and severe intrahepatic cholestasis after liver transplantation," Liver Transplantation, vol. 12, no. 11, pp. 1626–1633, 2006.
73. L. W. Teperman and V. P. Peyregne, "Considerations on the impact of hepatic encephalopathy treatments in the pretransplant setting," Transplantation, vol. 89, no. 7, pp. 771–778, 2010.
74. B. W. Shaw, R. P. Wood, R. D. Gordon, S. Iwatsuki, W. P. Gillquist, and T. E. Starzl, "Influence of selected patient variables and operative blood loss on six-month survival following liver transplantation," Seminars in Liver Disease, vol. 5, no. 4, pp.

385–393, 1985.

75. O. Selberg, J. Böttcher, G. Tusch, R. Pichlmayr, E. Henkel, and M. J. Müller, "Identification of high- and low-risk patients before liver transplantation: a prospective cohort study of nutritional and metabolic parameters in 150 patients," Hepatology, vol. 25, no. 3, pp. 652–657, 1997.
76. C. Miki, K. Iriyama, A. D. Mayer et al., "Energy storage and cytokine response in patients undergoing liver transplantation," Cytokine, vol. 11, no. 3, pp. 244–248, 1999.
77. J. M. Hasse, L. S. Blue, G. U. Liepa et al., "Early enteral nutrition support in patients undergoing liver transplantation," Journal of Parenteral and Enteral Nutrition, vol. 19, no. 6, pp. 437–443, 1995.
78. J. Reilly, R. Mehta, L. Teperman et al., "Nutritional support after liver transplantation: a randomized prospective study," Journal of Parenteral and Enteral Nutrition, vol. 14, no. 4, pp. 386–391, 1990.
79. S. Sherlock, "Hepatic encephalopathy," in Diseases of the Liver and Biliary System, pp. 109–110, Blackwell Scientific, Oxford, UK, 8th edition, 1989.
80. A. Donaghy, "Issues of malnutrition and bone disease in patients with cirrhosis," Journal of Gastroenterology and Hepatology, vol. 17, no. 4, pp. 462–466, 2002.
81. J. K. Heyman, C. J. Whitfield, K. E. Brock, G. W. McCaughan, and A. J. Donaghy, "Dietary protein intakes in patients with hepatic encephalopathy and cirrhosis: current practice in NSW and ACT,"Medical Journal of Australia, vol. 185, no. 10, pp. 542–543, 2006.
82. K. D. Mullen and S. Dasarathy, "Protein restriction in hepatic encephalopathy: necessary evil or illogical dogma?" Journal of Hepatology, vol. 41, no. 1, pp. 147–148, 2004.
83. M. Plauth, M. Merli, J. Kondrup, A. Weimann, P. Ferenci, and M. J. Müller, "ESPEN guidelines for nutrition in liver disease and transplantation," Clinical Nutrition, vol. 16, no. 2, pp. 43–55, 1997.
84. M. Plauth, M. Merli, and J. Kondrup, "Management of hepatic encephalopathy," New England Journal of Medicine, vol. 337,

no. 26, pp. 1921–1922, 1997.

85. M. Merli and O. Riggio, "Dietary and nutritional indications in hepatic encephalopathy," Metabolic Brain Disease, vol. 24, no. 1, pp. 211–221, 2009.

86. N. J. Greenberger, J. Carley, S. Schenker, I. Bettinger, C. Stamnes, and P. Beyer, "Effect of vegetable and animal protein diets in chronic hepatic encephalopathy," American Journal of Digestive Diseases, vol. 22, no. 10, pp. 845–855, 1977.

87. A. Keshavarzian, J. Meek, C. Sutton, V. M. Emery, E. A. Hughes, and H. J. Hodgson, "Dietary protein supplementation from vegetable sources in the management of chronic portal systemic encephalopathy," American Journal of Gastroenterology, vol. 79, no. 12, pp. 945–949, 1984.

88. G. P. Bianchi, G. Marchesini, A. Fabbri et al., "Vegetable versus animal protein diet in cirrhotic patients with chronic encephalopathy. A randomized cross-over comparison," Journal of Internal Medicine, vol. 233, no. 5, pp. 385–392, 1993.

89. D. K. Podolsky and K. J. Isselbacher, "Maladie alcoolique du foie et cirrhoses," in Harrison Médecine Interne, pp. 1483–1495, 13th edition, 1995.

90. S. Sherlock, Diseases of the Liver and Biliary System, Blackwell Scientific, 9th edition, 1993.

91. P. Felig and J. Wahren, "Protein turnover and amino acid metabolism in the regulation of gluconeogenesis," Federation Proceedings, vol. 33, no. 4, pp. 1092–1097, 1974.

92. P. B. Soeters and J. E. Fischer, "Insulin, glucagon, aminoacid imbalance, and hepatic encephalopathy,"Lancet, vol. 2, no. 7991, pp. 880–882, 1976.

93. T. Tomiya, Y. Inoue, M. Yanase et al., "Leucine stimulates the secretion of hepatocyte growth factor by hepatic stellate cells," Biochemical and Biophysical Research Communications, vol. 297, no. 5, pp. 1108–1111, 2002.

94. J. Córdoba, J. López-Hellín, M. Planas et al., "Normal protein diet for episodic hepatic encephalopathy: results of a randomized study," Journal of Hepatology, vol. 41, no. 1, pp. 38–43, 2004.

95. L. S. Eriksson, A. Persson, and J. Wahren, "Branched-chain amino acids in the treatment of chronic hepatic encephalopathy," Gut,

vol. 23, no. 10, pp. 801–806, 1982.

96. J. Wahren, J. Denis, P. Desurmont, et al., "Is intravenous administration of branched chain amino acids effective in the treatment of hepatic encephalopathy? A multicenter study," Hepatology, vol. 3, no. 4, pp. 475–480, 1983.

97. E. H. Egberts, H. Schomerus, W. Hamster, and P. Jürgens, "Branched chain amino acids in the treatment of latent portosystemic encephalopathy. A double-blind placebo-controlled crossover study,"Gastroenterology, vol. 88, no. 4, pp. 887–895, 1985.

98. M. Plauth, E. H. Egberts, W. Hamster et al., "Long-term treatment of latent portosystemic encephalopathy with branched-chain amino acids. A double-blind placebo-controlled crossover study,"Journal of Hepatology, vol. 17, no. 3, pp. 308–314, 1993.

99. G. Marchesini, G. Bianchi, M. Merli et al., "Nutritional supplementation with branched-chain amino acids in advanced cirrhosis: a double-blind, randomized trial," Gastroenterology, vol. 124, no. 7, pp. 1792–1801, 2003.

100. M. Iwasa, K. Matsumura, Y. Watanabe et al., "Improvement of regional cerebral blood flow after treatment with branched-chain amino acid solutions in patients with cirrhosis," European Journal of Gastroenterology and Hepatology, vol. 15, no. 7, pp. 733–737, 2003.

101. Y. Muto, S. Sato, A. Watanabe et al., "Effects of oral branched-chain amino acid granules on event-free survival in patients with liver cirrhosis," Clinical Gastroenterology and Hepatology, vol. 3, no. 7, pp. 705–713, 2005.

102. Y. Nakaya, K. Okita, K. Suzuki et al., "BCAA-enriched snack improves nutritional state of cirrhosis,"Nutrition, vol. 23, no. 2, pp. 113–120, 2007.

103. H. Fukushima, Y. Miwa, E. Ida et al., "Nocturnal branched-chain amino acid administration improves protein metabolism in patients with liver cirrhosis: comparison with daytime administration," Journal of Parenteral and Enteral Nutrition, vol. 27, no. 5, pp. 315–322, 2003.

104. G. Bianchi, R. Marzocchi, F. Agostini, and G. Marchesini, "Update on nutritional supplementation with branched-chain amino acids," Current Opinion in Clinical Nutrition and Metabolic Care, vol. 8, no. 1, pp. 83–87, 2005.

105. S. Khanna and S. Gopalan, "Role of branched-chain amino acids in liver disease: the evidence for and against," Current Opinion in Clinical Nutrition and Metabolic Care, vol. 10, no. 3, pp. 297–303, 2007.·

106. P. Clot, M. Tabone, S. Aricò, and E. Albano, "Monitoring oxidative damage in patients with liver cirrhosis and different daily alcohol intake," Gut, vol. 35, no. 11, pp. 1637–1643, 1994.

107. S. Moscarella, A. Duchini, and G. Buzzelli, "Lipoperoxidation, trace elements and vitamin E in patients with liver cirrhosis," European Journal of Gastroenterology and Hepatology, vol. 6, no. 7, pp. 633–636, 1994.

108. A. Nagita and M. Ando, "Assessment of hepatic vitamin E status in adult patients with liver disease,"Hepatology, vol. 26, no. 2, pp. 392–397, 1997.

109. S. A. Harrison, S. Torgerson, P. Hayashi, J. Ward, and S. Schenker, "Vitamin E and vitamin C treatment improves fibrosis in patients with nonalcoholic steatohepatitis," American Journal of Gastroenterology, vol. 98, no. 11, pp. 2485–2490, 2003.

110. M. Kugelmas, D. B. Hill, B. Vivian, L. Marsano, and C. J. McClain, "Cytokines and NASH: a pilot study of the effects of lifestyle modification and vitamin E," Hepatology, vol. 38, no. 2, pp. 413–419, 2003.

111. M. P. de la Maza, M. Petermann, D. Bunout, and S. Hirsch, "Effects of long-term vitamin E supplementation in alcoholic cirrhotics," Journal of the American College of Nutrition, vol. 14, no. 2, pp. 192–196, 1995.

112. S. Holt, D. Goodier, R. Marley et al., "Improvement in renal function in hepatorenal syndrome with N-acetylcysteine," Lancet, vol. 353, no. 9149, pp. 294–295, 1999.

113. G. Idéo, A. Bellobuono, S. Tempini et al., "Antioxidant drugs combined with alpha-interferon in chronic hepatitis C not responsive to alpha-interferon alone: a randomized, multicentre study,"European Journal of Gastroenterology and Hepatology, vol. 11, no. 11, pp. 1203–1207, 1999.

114. C. Bémeur, J. Vaquero, P. Desjardins, and R. F. Butterworth, "N-acetylcysteine attenuates cerebral complications of non-acetaminophen-induced acute liver failure in mice: antioxidant and anti-inflammatory mechanisms," Metabolic Brain Disease,

vol. 25, no. 2, pp. 241–249, 2010.

115. E. Cabré and M. A. Gassull, "Nutritional aspects of chronic liver disease," Clinical Nutrition, vol. 12, no. 1, pp. S52–S63, 1993.
116. S. Lévy, C. Hervé, E. Delacoux, and S. Erlinger, "Thiamine deficiency in hepatitis C virus and alcohol-related liver diseases," Digestive Diseases and Sciences, vol. 47, no. 3, pp. 543–548, 2002.
117. P. N. Newsome, I. Beldon, Y. Moussa et al., "Low serum retinol levels are associated with hepatocellular carcinoma in patients with chronic liver disease," Alimentary Pharmacology and Therapeutics, vol. 14, no. 10, pp. 1295–1301, 2000.
118. F. Gundling, N. Teich, H. M. Strebel, W. Schepp, and C. Pehl, "Nutrition in liver cirrhosis,"Medizinische Klinik, vol. 102, no. 6, pp. 435–444, 2007.
119. S. Masuda, T. Okano, K. Osawa, M. Shinjo, T. Suematsu, and T. Kobayashi, "Concentrations of vitamin D-binding protein and vitamin D metabolites in plasma of patients with liver cirrhosis," Journal of Nutritional Science and Vitaminology, vol. 35, no. 4, pp. 225–234, 1989.
120. W. A. Macbeth, E. N. Kass, and W. V. Mcdermott Jr., "Treatment of hepatic encephalopathy by alteration of intestinal flora with Lactobacillus acidophilus," The Lancet, vol. 285, no. 7382, pp. 399–403, 1965.
121. C. Loguercio, C. Del Vecchio Blanco, and M. Coltorti, "Enterococcus lactic acid bacteria strain SF68 and lactulose in hepatic encephalopathy: a controlled study," Journal of International Medical Research, vol. 15, no. 6, pp. 335–343, 1987.
122. C. Loguercio, R. Abbiati, M. Rinaldi, A. Romano, C. Del Vecchio Blanco, and M. Coltorti, "Long-term effects of Enterococcus faecium SF68 versus lactulose in the treatment of patients with cirrhosis and grade 1-2 hepatic encephalopathy," Journal of Hepatology, vol. 23, no. 1, pp. 39–46, 1995.
123. Q. Liu, Z. P. Duan, D. K. Ha, S. Bengmark, J. Kurtovic, and S. M. Riordan, "Synbiotic modulation of gut flora: effect on minimal hepatic encephalopathy in patients with cirrhosis," Hepatology, vol. 39, no. 5, pp. 1441–1449, 2004.

124. C. Loguercio, A. Federico, C. Tuccillo et al., "Beneficial effects of a probiotic VSL#3 on parameters of liver dysfunction in chronic liver diseases," Journal of Clinical Gastroenterology, vol. 39, no. 6, pp. 540–543, 2005.

125. M. Malaguarnera, F. Greco, G. Barone, M. P. Gargante, M. Malaguarnera, and M. A. Toscano, "Bifidobacterium longum with fructo-oligosaccharide (FOS) treatment in minimal hepatic encephalopathy: a randomized, double-blind, placebo-controlled study," Digestive Diseases and Sciences, vol. 52, no. 11, pp. 3259–3265, 2007.

126. M. Malaguarnera, M. P. Gargante, G. Malaguarnera et al., "Bifidobacterium combined with fructo-oligosaccharide versus lactulose in the treatment of patients with hepatic encephalopathy," European Journal of Gastroenterology and Hepatology, vol. 22, no. 2, pp. 199–206, 2010.

127. N. Rayes, D. Seehofer, S. Hansen et al., "Early enteral supply of Lactobacillus and fiber versus selective bowel decontamination: a controlled trial in liver transplant recipients," Transplantation, vol. 74, no. 1, pp. 123–128, 2002.

128. N. Rayes, D. Seehofer, T. Theruvath et al., "Supply of pre- and probiotics reduces bacterial infection rates after liver transplantation—a randomized, double-blind trial," American Journal of Transplantation, vol. 5, no. 1, pp. 125–130, 2005.

129. J. S. Bajaj, K. Saeian, K. M. Christensen et al., "Probiotic yogurt for the treatment of minimal hepatic encephalopathy," American Journal of Gastroenterology, vol. 103, no. 7, pp. 1707–1715, 2008.

Chapter 3

# Transgenic Vegetable Breeding for Nutritional Quality and Health Benefits

João Silva Dias[1] and Rodomiro Ortiz[2]

[1]Technical University of Lisbon, Instituto Superior de Agronomia, Tapada da Ajuda, Lisboa, Portugal

[2]Swedish University of Agricultural Sciences, Alnarp, Sweden

## ABSTRACT

Vegetables are essential for well-balanced diets. About 3 billion people in the world are malnourished due to imbalanced diets. Vegetables can contribute to the prevention of malnutrition disorders. Genetic engineering enables vegetable breeders to incorporate desired transgenes into elite cultivars, thereby improving their value considerably. It further offers unique opportunities for improving nutritional quality and bringing other health benefits. Many vegetable crops have been genetically modified to improve traits such as higher nutritional status or better flavour, and to reduce bitterness or anti-nutritional factors. Transgenic vegetables can be also used for vaccine delivery. Consumers could benefit further from eating more nutritious transgenic vegetables, e.g. an increase of crop carotenoids by metabolic sink manipulation through genetic engineering appears feasible in some vegetables. Genetically engineering carrots containing increase

Ca levels may boost Ca uptake, thereby reducing the incidence of Ca deficiencies such as osteoporosis. Fortified transgenic lettuce with zinc will overcome the deficiency of this micronutrient that severely impairs organ function. Folates deficiency, which is regarded as a global health problem, can also be overcome with transgenic tomatoes with folate levels that provide a complete adult daily requirement. Transgenic lettuce with improved tocopherol and resveratrol composition may prevent coronary disease and arteriosclerosis and can contribute to cancer chemopreventative activity. Food safety and health benefits can also be enhanced through transgenic approaches, e.g. rural African resource-poor consumers will benefit eating cyanide-free cultivars of cassava. Biotechnology-derived vegetable crops will succeed if clear advantages and safety are demonstrated to both growers and consumers.

# INTRODUCTION

Vegetables make up a major percentage of the human diet in many locations of the world and play a significant role in human nutrition, especially as sources of phytonutriceuticals: vitamins, minerals, dietary fiber and phytochemicals [1-3]. Some phytochemicals of vegetables are strong antioxidants and are thought to reduce the risk of chronic disease by protecting against free-radical damage, by modifying metabolic activation and detoxification of carcinogens, or even influencing processes that alter the course of tumor cells [1,2,4,5].

## Addressing Malnourishment through Improving Human Diets with Vegetables

About 3 billion people in the world are manourished due to imbalanced diets [6,7]. Underconsumption of vegetables and fruits is among the top 10 risk factors leading to micronutrient malnutrition and is associated with the prevalence of chronic diseases [8,9]. More than 70% of malnourished children live in Asia. At least half of the preschool children and pregnant women are affected by micronutrient deficiencies in Bangladesh, Cambodia, Nepal and the Philippines [10]. Deficiency of vitamin A and iron can result in severe anemia, impaired intellectual development, blindness, and even death.

Vegetables contain a range of macroand micro-nutrients, including pro-vitamin A, iron and zinc, which contribute to the prevention of malnutrition disorders. Small variation in maternal diets, particularly reduction in micronutrient content, can have a significant impact on fetal growth and development. Pregnant women, in particular, benefit from good vegetable nutrition, particularly during later pregnancy, and lactation. The interplay of the different micronutrients and antioxidants found in vegetables has important health impacts, explaining for instance the higher birth weight of children in India, when mothers consume higher rates of green leafy vegetables and fruits during pregnancy [11]. Nutrition is both a quantity and a quality issue, and vegetables in all their many forms ensure an adequate intake of most vitamins and nutrients, dietary fibers, and phytochemicals, which can bring muchneeded balanced diets thereby contributing to solve many of these nutritional problems.

## Biotecnology and Plant Breeding

Biotechnology is a new, and potentially powerful, tool that has been added by most of the multinational private seed sector to their vegetable breeding programs. It can augment or accelerate conventional cultivar development programs through saving time, delivering better products, and ensuring genetic uniformity, or achieving some outputs that are not possible by conventional breeding. Transgenic crops, commonly referred to as genetically modified (GM) crops enable plant breeders to bring favorable genes, often previously inaccessible, into elite cultivars, improving their value considerably and offer unique opportunities for controlling insects, viruses and other pathogens, as well as nutritional quality and health benefits. Conventional plant breeding that utilizes non-transgenic approaches will remain the backbone of vegetable genetic improvement strategies. However, transgenic crop cultivars should not be excluded as products capable of contributing to more nutritious and healthy food [12]. Many vegetable crops have been genetically modified to include resistance to pests, pathogens and herbicides, and for improved features, such as slow ripening, higher nutritional status, seedless fruit, and increased sweetness.

Recently, Dias and Ortiz [13] did an analysis of the status (until 2010) of transgenic tomato, eggplant, potato, cucurbits, brassicas, lettuce, alliums, sweet corn, cowpea, cassava, sweet potato, and carrots.

From this analysis we noticed several benefits of transgenic vegetables to nutritional quality and health benefits, which are very important for consumers. This article discusses recent attempts to characterize and modify phytonutriceuticals in vegetable crops by using transgenic approaches, focusing on those modifications that are of interest for improving human diets and health.

# TRANSGENIC BREEDING AND NUTRITIONAL HEALTH BENEFITS

## Vitamins and Carotenoids

Vitamin A is a major public health problem in much of the developing world. It is estimated that in developed countries, almost 70% of vitamin A comes from animal sources while 30% is derived from plant-based foods. In contrast, people in developing countries derive about 70% to 80% of vitamin A from plant-based foods. Vegetarians and populations with limited access to animal products depend on provitamin A carotenoids. Vitamin A deficiency affects approximately 25% of the developing world's preschoolers. It is associated with blindness, susceptibility to disease and higher mortality rates. It leads to the death of approximately 1 to 3 million children each year [14,15]. The most famous attempt to combat vitamin A deficiency is the development of GM "Golden Rice". Ye, et al. [16] genetically transformed rice genotypes with carotenoid biosynthetic genes to deliver more vitamin-A precursors in the diet. Although "Golden Rice" has not yet been commercialized [17], it has shown the potential for genetic manipulation of carotenoid biosynthesis in crops [18]. One of the most obvious benefits of enhancing carotenoid levels is the increase in pigmentation, which can lead to more deeply coloured vegetables that are often preferred by consumers. Thus, increasing levels of carotenoid is doubly beneficial, both in terms of nutrition and aesthetics. There are a range of other approaches to enhance the carotenoid levels in potatoes and other root vegetables. Diretto, et al. [19], have silenced the first step in the beta-epsilon branch of carotenoid biosynthesis, lycopene epsilon cyclase (LCY-e) in potato—a tuber crop that contains low levels of carotenoids. This antisense tuber-specific silencing of the gene results in significant increases in carotenoid levels, with up to 14-fold more β-carotene. Cervantes-Flores, et al. [20] have also recently reported in sweet potato the identification of quantitative trait loci (QTL) for dry matter, starch content and β-carotene content, opening up the possibility of genetic manipulation and further enhancement of this root crop.

Fruit-derived carotenoids might have greater health potential than root carotenoids because they have better bioavailability. Modification of carotenoid-containing fruit crops such as tomato and pepper might therefore have better outcomes than modification of root and tuber crops. Tomato fruit and its processed products are the principal dietary sources of carotenoids such as lycopene and β-carotene. Therefore, there is considerable interest in elevating the levels of carotenoids in tomato fruit by genetic manipulation and thereby improving the nutritional quality of the crop. To enhance the carotenoid content and profile of tomato fruit, Romer, et al. [21] produced transgenic lines containing a bacterial carotenoid gene (crtl) encoding the enzyme phytoene desaturase, which converts phytoene into lycopene. Expression of this gene in transgenic tomato plants of the cultivar "Ailsa Cray" did not elevate total carotenoid levels. However, the β-carotene content increased about threefold, up to 45% of the total carotenoid content. The alteration in carotenoid content of these transgenic tomatoes did not affect growth and development. Levels of noncarotenoid isoprenoids were unchanged in the transformants. The phenotype has been found to be stable and reproducible over at least four generations. Fraser, et al. [22] reported also an increase in tomato fruit carotenoids phytoene, lycopene, β-carotene and lutein in cultivar "Ailsa Craig". Phytoene synthase from the bacterium Erwinia uredovora (crtB) has been overexpressed in tomato cultivar "Ailsa Craig". Fruitspecific expression was achieved by using the tomato polygalacturonase promoter, and the CRTB protein was targeted to the chromoplast by the tomato phytoene synthase-1 transit sequence. Total fruit carotenoids of primary transformants (T(0)) were 2-4-fold higher than the controls, whereas phytoene, lycopene, β-carotene, and lutein levels were increased 2.4-, 1.8-, and 2.2-fold, respectively. The biosynthetically related isoprenoids, tocopherols plastoquinone and ubiquinone, were unaffected by changes in carotenoid levels. The progeny (T(1) and T(2) generations) inherited both the transgene and phenotype. Ripe tomato fruits accumulate large amounts of lycopene and small amounts of β-carotene (pro-vitamin A). Lycopene is transformed into β-carotene by the action of lycopene beta-cyclase (beta-Lcy). Rosati, et al. [23] introduced, via Agrobacterium-mediated transformation, DNA constructs aimed at up-regulating (OE construct) or down-regulating (AS construct) the expression of the beta-Lcy gene in a fruit-specific fashion. Three tomato transformants containing the OE construct show a significant increase in tomato fruit β-carotene content. The tomato fruits from these plants display different colour phenotypes, from orange to orange-red, depending on the lycopene/beta-carotene ratio. Fruits from AS transformants show up to 50% inhibition of beta-Lcy expression, accompanied by a slight increase in lycopene content. Leaf carotenoid composition is unaltered in all transformants. In most transformants, an increase in total carotenoid content is observed with respect to the parental line. This increase occurs in the absence of major variations in the expression of endogenous carotenoid genes.

Current advances in genetic engineering of brassicas have enable also the production of plants with alterations in a range of vitamins or amino acids for improved human nutrition. Lu, et al. [24] developed transgenic cauliflower with high levels of β-carotene accumulation, and Wahlroos, et al. [25] produced oilseed Brassica rapa with increased histidine content. It is likely in the future that more transgenic vegetable brassicas with altered vitamin or amino acid content will be also developed.

Vitamin E, which includes tocopherols, is a lipid soluble antioxidant. There are α, β, γ, and δ isoforms of tocopherol with relative vitamin E potencies of 100%, 50%, 10%, and 3%, respectively. Conversion of γ-tocopherol to α-tocopherol in vegetable crops could increase their value and importance in human health because vitamin E reduces the risk of several serious disorders (e.g. cardiovascular diseases and cancer), slows ageing and enhances the function of the immune system. Cho et al. [26] developed transgenic lettuce plants of the cultivar "Chungchima" expressing a cDNA encoding γ-tocopherol methyltransferase from Arabidopsis thaliana to improve tocopherol composition. Transgene inheritance and expression in transformed plants increased enzyme activity and conversion of γ-tocopherol to the more potent α form.

## Folates

Folate deficiency, regarded as a global health problem, causes neural tube defects and other human diseases. Folates are synthesized from pteridine, p-aminobenzoate (PABA), and glutamate precursos. Díaz de la Garza, et al. [27,28] developed trangenic tomatoes by engineering fruitspecific overexpression of GTP cyclohydrolase I that catalyzes the first step of pteridine synthesis, and aminodeoxychorismate synthase that catalyzes the first step of PABA synthesis. Vine-ripened fruits contained on average 25-fold more folate than controls by combining PABAand pteridine overproduction traits through crossbreeding of transgenic tomato plants. The achieved folate level provides a complete adult daily requirement with less than one standard serving. This strategy could have substantial implications for the biosynthesis of folate in vegetables that already produce this vitamin and for the general fortification of vegetables.

## Calcium and Zinc

Vegetables offer consumers a diverse mixture of nutrients that promote human health more beneficially than dietary supplements. However, essential nutrients such as calcium (Ca) could be limiting in plant-based diets. Consequently, genetically engineering vegetables containing increased Ca levels may boost Ca uptake, thereby reducing the incidence of Ca deficiencies such as osteoporosis. In this regard, Park, et al. [29] modified carrots to express increased levels of the plant Ca transporter sCAX1. These carrot lines were fertile and displayed no adverse phenotypes. Further mice and human feeding trials demonstrated increased Ca absorption from sCAX1-expressing transgenic carrots compared to controls [30]. This research supports alternative means of biofortifying vegetables with bioavailable Ca. Zinc is also an essential element in human nutrition, as its deficiency severely impairs organ function. It is also important for development of intellectual capacity. In experiments to fortify lettuce with this element, Zuo, et al. [31] used Agrobacterium-mediated gene delivery of a mouse metallothionein mutant β-cDNA in the lettuce cultivar "Salinas 88". The concentration of zinc in the lettuce transgenic plants increased to 400 μg/g dry weight, which is considerably higher than in wildtype plants.

## Phosphorous

Humans and animals also need to consume sufficient dietary phosphorus. Plant seeds usually contain an adequate amount of phosphorus, but most of it is in the form of phytic acid (inositol hexaphosphate)—i.e.; the storage form of phosphorus—which cannot be digested by nonruminants such as humans. Humans therefore excrete this excess phosphorus in their waste. Furthermore, phytic acid can act as metal chelator of important minerals such as calcium, magnesium, iron, zinc and can contribute to mineral deficiencies in people. About 20 years ago, researchers identified the first mutants in plants including corn, rice, wheat, soybean, barley and Arabidopsis that have lower amounts of phytic acid in their seeds and an increased amount of phosphorus. Low-phytic-acid germplasm accessions have been identified in lentil [32] and common bean [33], which extends this potential benefit to vegetable legumes. In addition to improving the phosphorus cycle, low-phytic-acid mutants provide further nutritional benefits, increasing the availability of minerals such as calcium, magnesium, iron, and zinc as these are no longer chelated by phytic acid. However, reducing phytic acid has resulted in a concomitant reduction in seed and plant performance; it compromises germination, emergence, stress tolerance and yield [34]. Several interesting biotechnological approaches have been suggested to remedy this problem, including embryo-specific silencing of an ABC transporter responsible for phytic acid accumulation [35] and the engineering of high-phytase seeds [36]. Biotechnological solutions might ultimately help to avoid phytic acid accumulation in seeds, but scientists have already demonstrated that mutagenesis was sufficient to develop a low-phytic acid mutant with yield performance comparable with its wild counterpart [37].

## Flavonoids

Flavonoids are polyphenols whose dietary intake has the potential to prevent chronic diseases. Schijlen, et al. [38] introduced heterologous, flavonoid pathway genes—stilbene synthase, chalcone synthase, chalcone reductase, chalcone isomerase, and flavone synthase—to produce novel flavonoids in tomato fruit. These novel flavonoidsflavones and flavonols-increased threefold, mostly in the Q12 peel, which had higher total antioxidant capacity. These findings add further support to the potential of engineering tomato fruit for accumulation of high levels of beneficial nutrients. Similarly, the poliphenol resveratrol—a stilbenoid—shows cancer chemo-preventative activity and may prevent coronary heart disease and arteriosclerosis. Liu, et al. [39] showed, in a quantitative analysis, that resveratrol in transgenic lettuce plants reached 56.0 ± 5.52 μ/g leaf fresh weight, which is comparable to that in the skin of grape fruit (Citrus × paridisi Macfad.). Flavonoids such as anthocyanins are known as antioxidants in vitro and can reduce the risk of many diseases related to aging. However, some vegetable brassicas, such as cauliflower, are low in anthocyanins. In an attempt to manipulate pigment biosynthesis to increase the health benefits of brassica vegetables, the effect of a regulatory locus of flavonoid content was assessed. Agrobacterium tumefaciens—mediated transformation of a B. oleracea line, selected for high transformation ability by Sparrow, et al. [40], was used to produce plants transgenic for the maize Lc (leaf color) locus. Lc is a regulatory gene in the anthocyanin pathway, and it is expected that its presence will increase the flavonoid content. Seedling explants were cocultivated with Agrobacterium tumefaciens strain LBA4404 containing a binary vector Q27 with a neomycin phosphotransferase II (NPTII) gene. Under tissue culture conditions, Lc-containing plants were green with no visible increase in anthocyanin production. However, after transfer to the greenhouse, the exposure to high light intensity led to visible signs of pigmentation within one week. Increased pigmentation was apparent in stems, petioles, main leaf veins, and sepals. Lc-containing lines had 10 to 20 times higher levels of total anthocyanins than controls. In addition, antioxidant activity of Lc-containing lines was 1.5 times higher than that of controls [41]. Potato does not normally produce anthocyanin, but germplasm expressing anthocyanin pigment has been developed and is attracting interest from consumers. Stushnoff, et al. [42] in potato identified 27 genes that are differentially expressed in purple and white tuber tissues. One of these genes—which encodes a novel single-domain MYB transcription factor—has the potential to influence anthocyanin-pigment production in potato. The resulting purple potato might offer both novelty and health functionality to consumers, who can also benefit from native Andean potatoes that do not always show desired tuber shapes for both table and processing industry.

## Organosulphur Compounds: Glucosinolates and Thiosulfides

Several epidemiological studies in Asia, the USA and Europe have suggested that the consumption of vegetables from the Brassicaceae family, notably broccoli, reduce the risk of lung, breast, colon, and prostate cancer [43]. The phytochemicals thought to be responsible for these health benefits are the isothiocyanates sulphoraphane and indole-3-carbinol. Sulphoraphane was initially thought to induce phase II enzymes in humans, which act against potentially carcinogenic compounds entering the body through the digestive system. However, it is not clear to what degree the anti-carcinogenic activity of sulphoraphane is achieved by phase II enzyme induction; it also seems that sulphoraphane can induce apoptosis and cell-cycle arrest in a variety of cell types [44]. While research continues into the health-promoting mechanisms of Brassica isothiocyanates, others have developed highglucosinolate broccoli germplasm that results in plants that produce mainly isothiocyanates, compared with standard broccoli cultivars that also produce nitriles [45]. Chromosome segments from a wild ancestor, Brassica villosa, have been introgressed to enhance glucosinolate levels. B. villosa alleles determine whether hydrolysis generates indole-3-carbinol or sulphoraphane. Hence, highglucosinolate broccoli might be suitable for increasing the amount of sulphoraphane in the diet. The extent to which vegetable brassicas protect against cancer probably depends on the genotype of the consumer, in particular the allele present at the GSTM1 locus. This gene codes for the enzyme glutathione transferase, which catalyses the conjugation of glutathione with isothiocyanates. Approximately 50% of humans carry a deletion of the GSTM1 gene [43], which reduces their ability to conjugate, process and excrete isothiocyanates. Individuals with two null alleles for GSTM1 might gain less protection from these cultivars of vegetable. The most commonly consumed Brassica vegetable in Asia is Brassica rapa. B. rapa contains different isothiocyanates to B. oleracea and recent evidence suggests that individuals who are null for GSTM1 can gain a protective benefit from B. rapa [46]. This example illustrates another aspect of complexity in breeding for health functionality in vegetable crops: human genetic variability has not generally been considered in the context of plant breeding programmes, but it might have important implications. Thus, when establishing vegetable breeding targets, it is important to explore the extent to which human variability affects the bioavailability and processing of health-functional compounds and influences health outcomes for a particular commodity.

Vegetables of the Allium genus such as onion, garlic, leek and chive are among the oldest crops associated with health-related properties [47]. Some of these traits appear to be related to the concentration and activity of organosulphur compounds in these vegetables [48]. The unique flavor and odor of alliums is derived from the hydrolysis of organosulfur compounds, which produces pyruvate, ammonia, and volatile sulfur compounds [49]. This reaction is catalyzed by the enzyme alliinase, which is contained in vacuoles within cells and released upon disruption of the tissue [50]. Variations in the ratios of these volatile sulfur compounds are responsible for the difference in flavors and odors between Allium species [49]. Along with health and nutritional benefits associated with these compounds, these thiosulfides are also major contributors to the bitter taste of some onions [49,50]. Three sets of transgenic onion plants containing antisense alliinase gene constructs (a CaMV 35S-driven antisense root alliinase gene, a CaMV 35S-driven antisense bulb alliinase, and a bulb alliinase promoter-driven antisense bulb alliinase) have been recently produced [51]. Results from the antisense bulb alliinase lines have been much more encouraging, and three lines were produced with barely detectable bulb alliinase levels and activity. Progress has been confounded by the poor survival of transgenic plants. Transgenic hybrid onion seed from these transgenic lines has been developed by crossing a nontransgenic openpollinated parental line with a transgenic parental plant carrying a single transgene in the hemizygous state. Some resulting seed produced by the nontransgenic parents will be hemizygous for the transgene and can be selected to obtain $F_1$ heterozygous individuals containing the transgene. Self-fertilization of these individuals produces homozygous, hemizygous, and null $F_2$ progeny for the transgene locus. These homozygous individuals can then be used to generate the bulk seed required for the production of commercial transgenic onion lines with less bitter taste.

When onions are cut, two compounds are formed: propanethial sulphoxide—also known as the lachrymatory factor—and 1-propanesulphenic acid. The lachrymatory factor reacts with nerve-cell membranes in the eye to produce tears, causing the familiar crying when cutting onions. In normal conditions, levels of 1-propanesulphenic acid are low because it is rapidly converted to the lachrymatory factor. Recently, Eady, et al. [52] silenced the gene for the lachrymatory factor enzyme by using RNAinterference, to produce tearless onions: 1-propanesulphenic acid self-condenses to 1-propenyl 1-propenethiosulphinate, which then undergoes further reactions [47]. This feat of genetic engineering reduces levels of lachrymatory factor up to 30-fold but does not diminish the overall levels of organosulphur compounds in the bulb. These "tearless onions" have potential health benefits for consumers as they do not produce tears, but retain their health-promoting properties.

## Sweet Taste

In attempts to reduce bitterness in lettuce, Sun, et al. [53] cloned the gene for the sweet and taste modifying protein miraculin from the pulp of berries of Richadella dulcifica, which is a West African shrub. This gene, with the CaMV 35S promoter, was introduced into the lettuce cultivar "Kaiser" using A. tumefaciens GV2260. Expression of this gene in transgenic plants led to the accumulation of significant concentrations of the sweet enhancing protein. People suffering diabetes may use miraculin, which is active at extremely low concentrations, as a food sweetener. The first successful study conducted to engineer genetically the taste of tomato fruit involved transformation of tomato with the thaumatin gene from the African plant katemfe (Thaumatococcus daniellii) [54]. Thaumatin is a sweet-tasting protein. Fruit from $T_2$ transgenic plants tasted sweeter than the control plants, leaving a unique and sweet-specific after taste.

Fructans and fructose polymers, sometimes known as inulin, might also have health-functional properties because they promote the growth of beneficial microbes in the gut, add sweetness without adding calories, and contribute to the fibre content of foods. Hellwege, et al. [55], developed transgenic potato plants that produce inulin by the expression of the 1-SST (sucrose:sucrose 1-fructosyltransferase) and 1-FFT (fructan:fructan 1-fructosyltransferase) genes from globe artichoke. The results suggested that these enzymes might be sufficient to produce inulin molecules of various lengths in plants.

## Anti-Nutritional Factors

Another approach to improving the health functionality of vegetable crops is to reduce the concentration of antinutritional factors. These are naturally occurring compounds with inhibitory effects on the nutritive potential of plants. In many cases, anti-nutritional factors are produced in planta for pest control, but have secondary effects on human nutrition. The first transgenic cassava plants became available in the mid-1990s [56-58] as plants with reduced cyanogenic content [59-61], which can benefit resource-poor people in rural Africa where this starchy root crop is the base of their diet. Faba bean (Vicia faba) contains condensed tannins that reduce the value of the inherently high protein levels of the crop. Tannins can be removed by the activity of two genes, zt-1 and zt-2, which are pleiotropic for white-flowered plants. Gutierrez, et al. [62] have identified a sequence characterized amplified repeat (SCAR) marker linked to the zt-2 gene that is associated with increased protein levels and reduced fibre content of faba bean seeds, which should facilitate the development of tannin-free faba cultivars. Calcium oxalate is another common anti-nutritional factor in plants. It is most commonly found as deposits in the vacuole of specialized cells called idioblasts [63]. The specific function of calcium oxalate accumulation in plants is not known; it might have a role in calcium regulation, ion balance, plant protection, detoxification or light gathering. There have been several attempts to reduce the amount of calcium oxalate in plant tissues by using molecular approaches. Nakata and McConn [64] identified mutants of barrel clover (Medicago truncatula) that are deficient in calcium oxalate and not compromised in growth. This suggests that it might be possible to genetically engineer plants with low or very low calcium oxalate levels; however, if calcium oxalate has a role in plant protection, low-calcium oxalate crops would require other protection strategies.

## Vaccine Delivery

Some vegetables, mainly tomato, have also been genetically modified to be used as vaccine delivery. Plant delivery of oral vaccines has attracted much attention because this strategy offers several advantages over vaccine delivery by injection [65-67]. Oral vaccines also offer the hope of more convenient immunization strategies and a more practical means of implementing universal vaccination programs worldwide. Tomato has been tested for expression of vaccines that can address human health issues of the developing world. Transgenic tomato plants potentially can bring several positive effects and improve human health. McGarvey, et al. [68] engineered tomato plants of cultivar "UC82b" to express a gene encoding a glycoprotein (G-protein), which coats the outer surface of the rabies virus. The recombinant constructs contained the G-protein gene from the environmental risk assessment strain of rabies virus. The G-protein was expressed in leaves and fruit of the transgenic plants, and it was found localized in Golgi bodies, vesicles, lasmalemma, and cell walls of vascular parenchyma cells. Ma, et al. [69] overexpressed hepatitis E virus (HEV) open reading frame 2 partial gene in tomato plants, to investigate its expression in transformants, the immunoactivity of expressed products, and explore the feasibility of developing a new type of plant-derived HEV oral vaccine. The recombinant protein was produced at 61.22 ng/g fresh weight in tomato fruits and 6.37 to 47.9 ng/g fresh weight in the leaves of the transformants. It was concluded that the HEV-E2 gene was correctly expressed in transgenic tomatoes and that the recombinant antigen derived had normal immunoactivity. These transgenic tomato plants are valuable tools for the development of edible oral vaccines. Chen, et al. [70] developed an effective antiviral agent against enterovirus 71 (EV71), which causes seasonal epidemics of hand, foot, and mouth disease associated with fatal neurological complications in young children, by transforming the gene for VP1 protein—a previously defined epitope and also a coat protein of EV71—in tomato plant. VP1 protein was first fused with sorting signals to enable it to be retained in the endoplasmic reticulum of tomato plant, and its expression level increased to 27 mg/g in fresh tomato fruit. Transgenic tomato fruit expressing VP1 protein was then used as an oral vaccine, and the development of VP1-specific fecal IgA and serum IgG were observed in BALB/c mice. Additionally, serum from mice fed transgenic tomato could neutralize the infection of EV71 to rhabdomyosarcoma cells, indicating that tomato fruit expressing VP1was successful in orally immunizing mice. Moreover, the proliferation of spleen cells from orally immunized mice was stimulated by VP1 protein and provided further evidence of both humoral and cellular immunity. Results of this study not only demonstrated the feasibility of using transgenic tomato as an oral vaccine to generate protective immunity in mice against EV71 but also the probability of enterovirus vaccine development. The Gram-negative bacterium Yersinia pestis causes plague, which has affected human health since ancient times. It is still endemic in Africa, Asia, and the American continent. There is the urgent need for a safe and cheap vaccine due to the increasing reports of the incidence of antibiotic-resistant strains and concern with the use of Y. pestis as an agent of biological warfare. Out of all the Y. pestis antigens tested, only $F_1$ and Vinduce a good protective immune response against a challenge with the bacterium [71]. Alvarez, et al. [72] reported the expression in tomato of the Y. pestis $F_1$-Vantigen fusion protein. The immunogenicity of the $F_1$-V transgenic tomatoes was confirmed in mice that were injected subcutaneously with bacterially produced $F_1$-V fusion protein and boosted orally with transgenic tomato fruit. Expression of the plague antigens in the tomato fruit allowed producing an oral vaccine candidate without protein purification and with minimal processing technology, offering a good system for a largescale vaccination programs in developing countries. The future of edible plant-based vaccines through transgenic approaches will depend on producing them safely on sufficient amounts.

## CONSTRAINTS FOR TRANSGENIC VEGETABLE BREEDING

As described consumers could benefit from improved access to more nutritious transgenic vegetables and enhanced food safety through transgenic approaches. However, transgenic crop technology for horticulture remains still in its infancy for several reasons. Vegetables are considered minor crops. Consequently, fewer resources are allocated to transgenic research of horticultural crops compared to field crops, especially by the multinational private seed corporations. While it is becoming less expensive to create transgenic crops, developing a marketable product and responding to the regulatory requirements remains very costly. Development and regulatory costs can be recouped more readily if the product is grown on an extensive area, which is generally not the case for individual vegetable crops. For this reason, most large multinational seed corporations have abandoned the development of transgenic vegetable crops.

Generally there are many cultivars of the same vegetable species on the market and the life span of an individual cultivar can be quite short. Introducing a transgene into a breeding program can be complicated and cost prohibitive, especially in crops with difficulty for using backcrossing (e.g. cassava, potato or sweet potato). Deregulation of a transgenic trait is event specific in many countries. For many vegetable species it is not possible to develop a single transgenic event that can be converted into many different cultivars of a single or closely related group of vegetable species through conventional breeding. For example, Brassica contains about 40 closely related commercialized crops, including cabbage (B. oleracea var. capitata), cauliflower (B. oleracea var. botrytis), broccoli (B. oleracea var. italica), Brussels sprouts (B. oleracea var. gemmifera), turnip (B. rapa var. rapa), broccoletto (B. rapa ssp. utilis), Chinese cabbage (B. rapa ssp. pekinensis), pak-choy (B. rapa ssp. chinensis), choysum (B. rapa ssp. parachinensis), swede or rutabaga (B. napus ssp. napobrassica), vegetable rape (B. napus ssp. napus), and various mustards (B. juncea, B. carinata and B. nigra) [73]. No single parent exists that can be used to backcross the transgene into the many different types of Brassica botanical varieties and subspecies. Individual events would have to be developed for many of the crop types, and deregulation of more than one event for a single protein is problematic for most business models. Because of the regulatory costs currently involved with GM vegetable crops, it is difficult for either the public or private sector to develop novel products specifically for small vegetable markets, including specialty vegetable crops in the developed and developing world and almost any crop in countries with relatively small agricultural sectors. For the few transgenic vegetable crops that are being developed, novel or unconventional strategies have been employed to bring the crops to markets, e.g. private-public partnerships in which the private sector would focus on selling hybrids to higher-end growers while the public sector would focus on low-resource farmers.

## SAFETY OF TRANSGENIC CROPS

Although transgenic cultivars have proven to be a powerful tool for nutritional health benefits, many countries are still engaged in discussions about their potential food safety. Consumer antagonism has precluded many farmers and other end users from sharing the benefits that these crops provide. Transgenic crops must pass a rigorous assessment for potential risks based on scientific data. The objective of this appraisal is to determine whether the transgenic crop is as safe as its conventional counterpart without transgenic modification. For this purpose, scientific data have to be produced to demonstrate that transgenic plants are safe for the environment and do not impose any health hazard for the consumers. In the USA, the process of deregulation is an interactive process between the industry, government agencies and any other stakeholder that feels concerned, where the industry has to provide scientific evidence as requested by the government agency to prove that there is no reasonable doubt on the safety of the transgenic crop.

The World Health Organization [74], the Food and Agriculture Organization of the United Nations [75], the Royal Society of London, the US National Academy of Sciences, the Brazilian Academy of Sciences, the Chinese Academy of Sciences, the Indian National Science Academy, the Mexican Academy of Sciences and the Third World Academy of Sciences [76], the American College of Nutrition [77], the Society of Toxicology [78], the British Medical Association [79], and the Union of German Academies of Sciences and Humanities [80], among others, have stated that GM crops approved for commercialization, do not pose more risk to human health than conventional crops, and they should be considered as safe as conventional ones.

The world has witnessed a steady increase of transgenic crop area in the last 1.5 decades. Extensive research has produced no evidence that transgenic crops approved by the authorities impose a greater risk to human and animal health than conventional crops. The Federal Office of Consumer Protection and Food Safety of Germany and partners [81] issued the report "Biological and Ecological Evaluation towards Long-term Effects" (also known as the BEETLE report) with the aim of providing scientific data to the European Commission. The BEETLE report reviewed in excess of 100 publications and consulted 52 experts in health issues to assess the possible long-term effect of GM crops on the health of consumers and the environment. This report concluded that so far no adverse effects to human health from eating GM food have been found. The report further stated that although unexpected negative effects are known in conventional crops, none has yet been detected in GM crops. The report concludes that there is a negligible probability for adverse effects to consumers' health in the long term.

# CONCLUSIONS

Transgenic cultivars can improve nutritional quality and health benefits and make important contributions to sustainable vegetable production by overcoming limiting factors in production, mainly virus diseases and pests, which are not easily addressed through conventional vegetable breeding alone. The variety of transgenic and conventional breeding techniques being used to enhance the nutritional quality and health benefits of vegetable crops is increasing. Molecular approaches hold great promise for future modifications. To be successful, however, more interdisciplinary work is required and besides molecular biologists and plant breeders is necessary to involve nutritional and food scientists as well as others from biomedical fields to ascertain the true function of specific plant compounds. A barrier to the successful use of transgenic techniques might be the acceptance—or lack thereof—of transgenic vegetable crops by the public. Strategies for improving the health functionality of vegetables that rely on transgenic approaches offer great scientific promise, but have so far been met with public scepticism, and even fear. Thus far, only two transgenic vegetable species—squash and sweet corn—has been approved and sold commercially for any length of time in the USA, and, despite the brief appearance and quick disappearance of transgenic tomatoes and potatoes in the past twenty years, there is widespread doubt whether more genetically modified vegetable crops will be approved in the near term. It remains to be seen what the marketplace will bring in the next decade. Fortunately for both scientists and consumers, increasing interest in this area has fuelled research, which bodes well for improving our understanding of the health functionality of vegetables and the potential for developing transgenic vegetable crops.

Transgenic vegetable crops are not a silver bullet for achieving nutritional quality and health benefits but coupled with conventional breeding can be a powerful tool for making available better vegetables. Biotechnologyderived vegetable crops will succeed if clear advantages and safety are demonstrated to both growers and consumers.

# REFERENCES

1. W. Craig and L. Beck, "Phytochemicals: Health Protective Effects," Canadian Journal of Dietetic Practice and Research, Vol. 60, 1999, pp. 78-84.
2. M. J. Wargovich, "Anticancer Properties of Fruits and Vegetables," HortScience, Vol. 35, 2000, pp. 573-575.
3. J. S. Dias and E. Ryder, "World Vegetable Industry: Production, Breeding, Trends," Horticulture Review, Vol. 38, 2011, pp. 299-356.
4. S. Southon, "Increased Fruit and Vegetable Consumption within the EU: Potential Health Benefits," Food Research International, Vol. 33, 2000, pp. 211-217. doi:10.1016/S0963-9969(00)00036-3
5. E. Herrera, R. Jimenez, O. I. Aruoma, S. Hercberg, I. Sanchez-Garcia and C. Fraga, "Aspects of Antioxidant Foods and Supplements in Health and Disease," Nutrition Reviews, Vol. 67, 2009, pp. S140-S144. doi:10.1111/j.1753-4887.2009.00177.x
6. W. Pfeiffer and B. McClafferty, "HarvestPlus: Breeding Crops for Better Nutrition," Crop Science, Vol. 57, 2007, pp. s88-s105.
7. J. D. H. Keatinge, F. Waliyar, R. H. Jammadass, A. Moustafa, M. Andrade, P. Drechsel, J. d'A. Hughes, P. Kardivel and K. Luther, "Re-Learning Old Lessons for the Future of Food: By Bread Alone No Longer—Diversifying Diets with Fruit and Vegetables," Crop Science, Vol. 50, 2010, pp. 51-62.
8. J. S. Dias, "World Importance, Marketing and Trading of Vegetables," Acta Horticulturae, Vol. 921, 2011, pp. 153- 169.
9. WHO (World Health Organization), "Diet, Nutrition and the Prevention of Chronic Diseases," Report of a Joint WHO/FAO Expert Consultation, WHO Technical Report Ser. 916 World Health Organization, Geneva, 2003.
10. Hellen Keller International, "Homestead Food Production Model Contributes to Improved Household Food Security, Nutrition and Female Empowerment—Experience from Scaling-Up Programs in Asia (Bangladesh, Cambodia, Nepal and Philippines)," Nutrition Bulletin, Vol. 8, 2010, pp. 1-8.

11. S. Rao, C. S. Yajnik, A. Kanade, D. H. D. Fall, B. M. Margetts and A. A. Jackson, "Intake of MicronutrientRich Foods in Rural Indian Mothers Is Associated with the Size of Their Babies at Birth: Pune Maternal Nutrition Study," Journal of Nutrition, Vol. 131, 2001, pp. 1217- 1224.
12. M. Newell-McGloughlin, "Nutritionally Improved Agricultural Crops," Plant Physiology, Vol. 147, 2008, pp. 939-953. doi:10.1104/pp.108.121947
13. J. S. Dias and R. Ortiz, "Transgenic Vegetable Crops: Progress, Potentials and Prospects," Plant Breeding Reviews, Vol. 35, 2012, pp. 151-246.
14. WFP (World Food Programme), "Annual Report," WFPUN, Washington, 2007.
15. UNICEF, "Vitamin and Mineral Deficiency. A Global Progress Report," UNICEF, 2010.
16. X. Ye, S. Al-Babili, A. Klöti, J. Zhang, P. Lucca, P. Beyer and I. Potrykus, "Engineering the Provitamin A (Beta-Carotene) Biosynthetic Pathway into (CarotenoidFree) Rice Endosperm," Science, Vol. 287, 2000, pp. 303- 305. doi:10.1126/science.287.5451.303
17. M. Enserink, "Tough Lessons from Golden Rice," Science, Vol. 320, 2008, pp. 468-471.doi:10.1126/science.320.5875.468
18. D. DellaPenna and B. Pogson, "Vitamin Synthesis in Plants: Tocopherols and Carotenoids," Annual Review of Plant Biology, Vol. 57, 2006, pp. 711-738.doi:10.1146/annurev.arplant.56.032604.144301
19. G. Diretto, R. Tavazza, R. Welsch, D. Pizzichini, F. Mourgues, V. Papacchioli, P. Beyer and G. Giuliano, "Metabolic Engineering of Potato Tuber Carotenoids through Tuber-Specific Silencing of Lycopene Epsilon Cyclase," BMC Plant Biology, Vol. 6, 2006, p. 13.doi:10.1186/1471-2229-6-13
20. J. C. Cervantes-Flores, B. Sosinski, K. V. Pecota, R. O. M. Mwanga, G. L. Catignani, V. D. Truong, R. H. Watkins, M. R. Ulmer and G. C. Yencho, "Identification of Quantitative Trait Loci for Dry-Matter, Starch, and Beta Carotene Content in Sweet Potato," Molecular Breeding, Vol. 28, 2010, pp. 201-216. doi:10.1007/s11032-010-9474-5

21. S. Romer, P. D. Fraser, J. W. Kiano, C. A. Shipton, N. Misawa, W. Schuch and P. M. Bramley, "Elevation of the Provitamin A Content of Transgenic Tomato Plants," Nature Biotechnology, Vol. 18, 2000, pp. 666-669. doi:10.1038/76523

22. P. D. Fraser, S. Romer, C. A. Shipton, P. B. Mills, K. W. Kiano, N. Misawa, R. G. Drake, W. Schuch and P. M. Bramley, "Evaluation of Transgenic Tomato Plants Expressing an Additional Phytoene Synthase in a Fruit Specific Manner," Proceedings of the National Academy of Sciences, Vol. 99, 2002, pp. 1092-1097. doi:10.1073/pnas.241374598

23. C. Rosati, R. Aquilani, S. Dharmapur, P. Pallara, C. Marusic, R. Tavazza, F. Bouvier, B. Camara and G. Giuliano, "Metabolic Engineering of Beta-Carotene and Lycopene Content in Tomato Fruit," Plant Journal, Vol. 24, 2000, pp. 413-419. doi:10.1046/j.1365-313x.2000.00880.x

24. S. Lu, J. van Eck, X. Zhou, A. B. Lopez, D. M. O'Halloran, K. M. Cosman, B. J. Conlin, D. J. Paolillo, D. F. Garvin, J. Vrebalov, L. V. Kochian, H. Kupper, E. D. Earle, J. Cao and L. Li, "The Cauliflower Or Gene Encodes a DnaJ Cysteine-Rich Domain-Containing Protein that Mediates High Levels of β-Carotene Accumulation," Plant Cell, Vol. 18, 2006, pp. 3594-3605. doi:10.1105/tpc.106.046417

25. T. Wahlroos, P. Susi, A. Solovyev, Y. Dorokhov, S. Morozov, J. Atabekov and T. Korpela, "Increase of Histidine Content in Brassica rapa subsp. oleifera by OverExpression of Histidine-Rich Fusion Proteins," Molecular Breeding, Vol. 14, 2004, pp. 455-462.doi:10.1007/s11032-004-0902-2

26. E. A. Cho, C. A. Lee, Y. S. Kim, S. H. Baek, B. G. Reyes and S. J. Yun, "Expression of g-tocopherol methyltransferase Transgene Improves Tocopherol Composition in Lettuce (Lactuca sativa L.)," Molecules & Cells, Vol. 19, 2005, pp. 16-22.

27. R. I. Diaz de la Garza, E. P. Quinlivan, S. M. J.. Klaus, G. J. C. Basset, J. F. Gregory and A. D. Hanson, "Folate Biofortification in Tomatoes by Engineering the Pteridine Branch of Folate Synthesis," Proceedings of National Academy of Sciences, Vol. 101, 2004, pp. 13720-13725. doi:10.1073/pnas.0404208101

28. R. I. Diaz de la Garza, J. F. Gregory and A. D. Hanson, "Folate Biofortification of Tomato Fruit," Proceedings of National

Academy of Sciences, Vol. 104, 2007, pp. 4218- 4222. doi:10.1073/pnas.0700409104

29. S. Park, C. K. Kim, L. M. Pike, R. H. Smith and K. D. Hirschi, "Increased Calcium in Carrots by Expression of an Arabidopsis $H^+/Ca^{2+}$ Transporter," Molecular Breeding, Vol. 14, 2004, pp. 275-282. doi:10.1023/B:MOLB.0000047773.20175.ae
30. J. Morris, K. M. Hawthorne, T. Hotze, S. A. Abrams and K. D. Hirschi, "Nutritional Impact of Elevated Calcium Transport Activity in Carrots," Proceedings of National Academy of Sciences, Vol. 105, 2008, pp. 1431-1435. doi:10.1073/pnas.0709005105
31. X. Zuo, Y. Zhang, B. Wu, X. Chang and B. Ru, "Expression of the Mouse Metallothionein Mutant ββ-cDNA in the Lettuces (Lactuca sativa L.)," Chinese Science Bulletin, Vol. 47, 2002, pp. 558-562. doi:10.1360/02tb9128
32. P. Thavarajah, D. Thavarajah and A. Vandenberg, "Low Phytic Acid Lentils (Lens culinaris L.): A Potential Solution for Increased Micronutrient Bioavailability," Journal of Agricultural and Food Chemistry, Vol. 57, 2009, pp. 9044-9049. doi:10.1021/jf901636p
33. M. Fileppi, I. Galasso, G. Tagliabue, M. G. Daminati, B. Campion, E. Doria and F. Sparvoli, "Characterisation of Structural Genes Involved in Phytic Acid Biosynthesis in Common Bean (Phaseolus vulgaris L.)," Molecular Breeding, Vol. 25, 2010, pp. 453-470. doi:10.1007/s11032-009-9344-1
34. V. Raboy, "The ABCs of Low-Phytate Crops," Nature Biotechnology, Vol. 25, 2007, pp. 874-875. doi:10.1038/nbt0807-874
35. J. Shi, H. Wang, K. Schellin, B. Li, M. Faller, J. M. Stoop, R. B. Meeley, D. S. Ertl, J. P. Ranch and K. Glassman, "Embryo-Specific Silencing of a Transporter Reduces Phytic Acid Content of Maize and Soybean Seeds," Nature Biotechnology, Vol. 25, 2007, pp. 930-937. doi:10.1038/nbt1322
36. V. Raboy, "Approaches and Challenges to Engineering Seed Phytate and Total Phosphorus," Plant Science, Vol. 177, 2009, pp. 281-296.doi:10.1016/j.plantsci.2009.06.012
37. B. Campion, F. Sparvoli, E. Doria, G. Tagliabue, I. Galasso, M. Fileppi, R. Bollini and E. Nielsen, "Isolation and Characterisation of an LPA (Low Phytic Acid) Mutant in Common Bean (Phaseolus vulgaris L.)," Theoretical and Applied Genetics, Vol. 118, 2009, pp. 1211- 1221. doi:10.1007/s00122-009-0975-8

38. E. Schijlen, C. H. Ric de Vos, H. Jonker, H. van den Broeck, J. Molthoff, A. van Tunen, S. Martens and A. Bovy, "Pathway Engineering for Healthy Phytochemicals Leading to the Production of Novel Flavonoids in Tomato Fruit," Plant Biotechnology Journal, Vol. 4, 2006, pp. 433-444. doi:10.1111/j.1467-7652.2006.00192.x
39. S. Liu, Y. Hu, X. Wang, J. Zhong and Z. Lin, "High Content of Resveratrol in Lettuce Transformed with a Stilbene Synthase Gene of Partenocissus henryana," Journal of Agricultural and Food Chemistry, Vol. 54, 2006, pp. 8082-8085. doi:10.1021/jf061462k
40. P. A. C. Sparrow, P. J. Dale and J. A. Irwin, "The Use of Phenotypic Markers to Identify Brassica oleracea Genotypes for Routine High-Throughput Agrobacterium Mediated Transformation," Plant Cell Reports, Vol. 23, 2004, pp. 64-70. doi:10.1007/s00299-004-0818-7
41. R. H. Braun, S. C. Morrison, K. E. Schwinn and M. C. Christey, "Agrobacterium mediated transformation of Brassica oleracea with the Lc locus," Abstracts International Association for Plant Tissue Culture and Biotechnology, Beijing, 13-18 August 2006, p. 115.
42. C. Stushnoff, L. J. Ducreux, R. D. Hancock, P. E. Hedley, D. G. Holm, G. J. McDougall, J. W. McNicol, J. Morris, W. L. Morris, J. A. Sungurtas, S. R. Verrall, T. Zuber and M. A. Taylor, "Flavonoid Profiling and Transcriptome Analysis Reveals New Gene-Metabolite Correlations in Tubers of Solanum tuberosum L.," Journal of Experimental Botany, Vol. 61, 2010, pp. 1225-1238.
43. N. Juge, R. F. Mithen and M. Traka, "Molecular Basis for Chemoprevention by Sulforaphane: A Comprehensive Review," Cellular and Molecular Life Science, Vol. 64, 2007, pp. 1105-1127.
44. S. Kumar and A. Andy, "Health Promoting Bioactive Phytochemicals from Brassica," International Food Research Journal, Vol. 19, 2012, pp. 141-152.
45. G. Sarikamis, J. Marquez, R. MacCormack, R. N. Bennett, J. Roberts and R. Mithen, "High Glucosinolate Broccoli: A Delivery System for Sulforaphane," Molecular Breeding, Vol. 18, 2006, pp. 219-228.

46. A. V. Gasper, A. Al-Janobi, J. A. Smith, J. R. Bacon, P. Fortun, C. Atherton, M. A Taylor, C. J. Hawkey, D. A. Barrett and R. F. Mithen, "Glutathione-S-Transferase M1 Polymorphism and Metabolism of Sulforaphane from Standard and High-Glucosinolate Broccoli," American Journal of Clinical Nutrition, Vol. 82, 2005, pp. 1283-1291.
47. E. Block, "Garlic and Other Alliums: The Lore and the Science," RSC, Cambridge, 2010.
48. P. Rose, M. Whiterman, P. K. Moore and Y. Z. Zhu, "Bioactive S-Alk(en)yl Cysteine Sulfoxide Metabolites in the Genus Allium: The Chemistry of Potential Therapeutic Agents," Natural Product Reports, Vol. 22, 2005, pp. 351-368.
49. W. M. Randle and J. E. Lancaster, "Sulphur Compounds in Alliums in Relation to Flavour Quality," In: J. L. Brewster, Ed., Onions and Other Vegetable Alliums, CAB International, Wallingford, 2002, pp. 329-356.
50. J. E. Lancaster and H. A. Collin, "Presence of Alliinase in Isolated Vacuoles and of Alkyl Cysteine Sulphoxides in the Cytoplasm of Bulbs of Onion (Allium cepa L.)," Plant Science Letters, Vol. 22, 1981, pp. 169-176.
51. C. C. Eady, S. Davis, J. Farrant, J. Reader and F. Kenel, "Agrobacterium tumefaciens Mediated Transformation and Regeneration of Herbicide Resistant Onion (Allium cepa) Plants," Annals of Applied Biology, Vol. 142, 2003, pp. 213-217.
52. C. C. Eady, T. Kamoi, M. Kato, N. G. Porter, S. Davis, M. Shaw, A. Kamoi and S. Imai, "Silencing Onion Lachrymatory Factor Synthase Causes a Significant Change in the Sulfur Secondary Metabolite Profile," Plant Physiology, Vol. 147, 2008, pp. 2096-2106.
53. H. J. Sun, M. L. Cui, B. Ma and H. Ezura, "Functional Expression of the Taste Modifying Protein, Miraculin, in Transgenic Lettuce," FEBS Letters, Vol. 580, 2006, pp. 620-626.
54. G. Bartoszewski, A. Niedziela, M. Szwacka and K. Niemirowicz-Szczytt, "Modification of Tomato Taste in Transgenic Plants Carrying a Thaumatin Gene from Thaumatococcus daniellii Bent.," Plant Breeding, Vol. 122, 2008, pp. 347-351.
55. E. M. Hellwege, S. Czapla, A. Jahnke, L. Willmitzer and A. G. Heyer, "Transgenic Potato (Solanum tuberosum) Tubers Synthesize the

Full Spectrum of Inulin Molecules Naturally Occurring in Globe Artichoke (Cynara scolymus) Roots," Proceedings of National Academy of Sciences, Vol. 97, 2000, pp. 8699-8704.

56. H. Q. Li, C. Sautter, I. Potrykus and J. Pounti-Kaerlas, "Genetic Transformation of Cassava (Manihot esculenta Crantz)," Nature Biotechnology, Vol. 14, 1996, pp. 736- 740.

57. C. J. J. M. Raemakers, E. Sofiari, N. J. Taylor, G. G. Henshaw, E. Jacobsen and R. G. F. Visser, "Production of Transgenic Cassava Plants by Particle Bombardment Using Luciferase Activity as the Selection Marker," Molecular Breeding, Vol. 2, 1996, pp. 339-349.

58. C. Schopke, N. J. Taylor, R. Carcamo, N. K. Konan, P. Marmey, G. G. Henshaw, R. N. Beachy and C. M. Fauquet, "Regeneration of Transgenic Cassava Plants (Manihot esculenta Crantz) from Microbombarded Embryogenic Suspension Cultures," Nature Biotechnology, Vol. 14, 1996, pp. 731-735.

59. D. Siritunga and R. T. Sayre, "Generation of CyanogenFree Transgenic Cassava," Planta, Vol. 217, 2003, pp. 367-373.

60. D. Siritunga and R. T. Sayre, "Engineering Cyanogen Synthesis and Turnover in Cassava (Manihot esculenta)," Plant Molecular Biology, Vol. 56, 2004, pp. 661-669.

61. D. Siritunga, D. Arias-Garzon, W. White and R. T. Sayre, "Over-Expression of Hydroxynitrile Lyase in Transgenic Cassava Roots Accelerates Cyanogenesis and Food Detoxification," Plant Biotechnology Journal, Vol. 2, 2004, pp. 37-43.

62. N. Gutierrez, C. M. Avila, M. T. Moreno and A. M. Torres, "Development of SCAR Markers Linked to zt-2, One of the Genes Controlling Absence of Tannins in Faba Bean," Australian Journal of Agricultural Research, Vol. 59, 2008, pp. 62-68.

63. P. A. Nakata, "Advances in Our Understanding of Calcium Oxalate Crystal Formation and Function in Plants," Plant Science, Vol. 164, 2003, pp. 901-909.

64. P. A. Nakata and M. M. McConn, "Calcium Oxalate Crystal Formation Is Not Essential for Growth of Medicago truncatula," Plant Physiology and Biochemistry, Vol. 41, 2003, pp. 325-329.

65. W. H. R. Langridge, "Edible Vaccines," Scientific American, 2000, pp. 66-71.

66. D. W. Pascual, "Vaccines Are for Dinner," Proceedings of National Academy of Sciences, Vol. 104, 2007, pp. 10757-10758.

67. G. B. Sunil Kumar, T. R. Gananpathi and V. A. Bapat, "Production of Hepatitis B Surface Antigen in Recombinant Plant Systems," Biotechnology Progress, Vol. 23, 2007, pp. 523-529.

68. P. B. McGarvey, J. Hammond, M. M. Dienelt, D. C. Hooper, Z. F. Fu, B. Dietzschold, H. Koprowski and F. H. Michaels, "Expression of the Rabies Virus Glycoprotein in Transgenic Tomatoes," Bio/ Technology, Vol. 13, 1995, pp. 1484-1487.

69. Y. Ma, J. Zhang, S. Q. Lin and N. S. Xia, "Genetic Engineering Vaccines Produced by Transgenic Plants," Xiamen Daxue Xuebao, Vol. 40, 2001, pp. 71-77.

70. H. F. Chen, M. H. Chang, B. L. Chiang and S. T. Jeng, "Oral Immunization of Mice Using Transgenic Tomato Fruit Expressing VP1 Protein from Enterovirus 71," Vaccine, Vol. 24, 2006, pp. 2944-2951.

71. G. Benner, G. Andrews, W. Byrne, S. Strachan, A. Sample and D. Heath, "Immune Response to Yersinia Outer Proteins and Other Yersinia pestis Antigens after Experimental Plague Infection in Mice," Infection and Immunity, Vol. 67, 1999, pp. 1922-1928.

72. M. L. Alvarez, H. L. Pinyerd, J. D. Crisantes, M. M. Rigano, J. Pinkhasov, M. Amanda, A. M. Walmsley, H. S. Masona and G. A. Cardineau, "Plant-Made Subunit Vaccine against Pneumonic and Bubonic Plague Is Orally Immunogenic in Mice," Vaccine, Vol. 24, 2006, pp. 2477-2490.

73. S. J. Kays and J. S. Dias, "Common Names of Commercially Cultivated Vegetables of the World in 15 Languages," Economic Botany, Vol. 49, 1995, pp. 115-152.

74. WHO (World Health Organization), 20 Preguntas Sobre Los Alimentos Modificados Genéticamente. WHO, Geneva, 2002. http://www.who.int/foodsafety/publications/biotech/en/20questions_es.pdf

75. FAO, FAO Statement on Biotechnology. Food and Agriculture Organization of the United Nations, Rome, Italy, 2000. http://www.fao.org/biotech/stat.asp?lang=en

76. Royal Society of London, the US National Academy of Sciences, the Brazilian Academy of Sciences, the Chinese Academy of

Sciences, the Indian National Science Academy, the Mexican Academy of Sciences and the Third World Academy of Sciences, Transgenic Plants and World Agriculture. National Academies Press, Washington DC, 2000.

77. B. M. Chassy, "Food Safety Evaluation of Crops Produced through Biotechnology," Journal of the American College of Nutrition, Vol. 21, 2002, pp. S166-S173.

78. Society of Toxicology, "The Safety of Genetically Modified Foods Produced through Biotechnology," Society of Toxicology, Virginia, 2002. http://www.toxicology.org/ai/gm/GM_Food.asp

79. British Medical Association, "Genetically Modified Foods and Health: A Second Interim Statement," British Medical Association, London, 2004. http://www.bma.org.uk/images/GM_tcm41-20804.pdf

80. Union of the German Academies of Science and Humanities, "Are there Health Hazards for the Consumers from Eating Genetically Modified Foods?" Interacademy Panel Initiative on Genetically Modified Organisms, Berlin, 2006. http://www.akademienunion.de/_files/memorandum_gentechnik/GMGeneFood.pdf

81. Federal Office of Consumer Protection and Food Safety (Germany) and Partners, "Long-Term Effects of Genetically Modified (GM) Crops on Health and the Environment (including Biodiversity): Prioritisation of Potential Risks and Delimitation of Uncertainties," Federal Office of Consumer Protection of Food Safety, Berlin, 2009. http://bch.cbd.int/database/record-v4-shtml?documentid=101007

Chapter 4

# Magnesium and Vascular Changes in Hypertension

Ana Rosa Cunha, Bianca Umbelino, Margarida L. Correia, and Mario Fritsch Neves

Department of Clinical Medicine, University Hospital Pedro Ernesto, State University of Rio de Janeiro, Avenida 28 de Setembro, 77 Sala 329, 20551-030, Rio de Janeiro, RJ, Brazil

## ABSTRACT

Many factors have been implicated in the pathogenesis of hypertension, including changes in intracellular concentrations of calcium, sodium, potassium, and magnesium. There is a significant inverse correlation between serum magnesium and incidence of cardiovascular diseases. Magnesium is a mineral with important functions in the body such as antiarrhythmic effect, actions in vascular tone, contractility, glucose metabolism, and insulin homeostasis. In addition, lower concentrations of magnesium are associated with oxidative stress, proinflammatory state, endothelial dysfunction, platelet aggregation, insulin resistance, and hyperglycemia. The conflicting results of studies evaluating the effects of magnesium supplements on blood pressure and other cardiovascular outcomes indicate that the action of magnesium in the vascular system is present but not yet established. Therefore, this mineral supplementation is not indicated as part of antihypertensive treatment, and further studies are needed to better cardiovascular diseases.

# INTRODUCTION

Primary hypertension is the most common form of blood pressure elevation whose cause remains unknown. However, many factors have been implicated in its pathogenesis, such as the renin-angiotensin-aldosterone system and the sympathetic nervous system hyperactivation. In addition, changes in intracellular ions such as calcium, sodium, potassium, and magnesium have also been related to high blood pressure.

In the last years, the prevalence of hypertension is around 25–30% in developed countries [1], and several treatments have been proposed for the BP control and prevention of its onset. Among the various studies concerning non-pharmacological treatments, there is need for lifestyle change with the inclusion of regular physical activity and healthy eating habits.

Observational studies have shown that a diet rich in potassium, magnesium, and calcium, present mainly in fruits and vegetables, is associated with lower incidence and mortality from cardiovascular disease [2]. In particular, magnesium has been the target of many studies [3], considering that there is a significant inverse correlation between serum magnesium levels and incidence of cardiovascular diseases [4]. In addition, hypertensive patients generally exhibit reduced intracellular concentrations of magnesium, while the contents of sodium and calcium are often increased compared to normotensive subjects [5, 6].

The dietary recommendation (Recommended Dietary Allowances/ RDA) for magnesium is 400 to 420 mg daily for adult men and 310 to 320 mg daily for adult women. However, consumption is far below this recommendation, and the high prevalence of this deficiency has been associated to several chronic diseases. Magnesium is found in most foods, but in varying concentrations. Leafy vegetables, nuts, whole grains, fruits, and legumes are considered as foods with high-magnesium concentrations [7].

In order to gather more information about the association of magnesium with cardiovascular diseases, we performed a narrative review of the literature through the PubMed database with the following descriptor: magnesium, intracellular magnesium, hypertension, arterial stiffness, and endothelial function. We included narrative reviews, experimental protocols, and controlled studies in the last 15 years (1996–2011), and case reports were excluded.

# PHYSIOLOGICAL FUNCTIONS AND PATHOPHYSIOLOGICAL ACTIONS OF MAGNESIUM

The mineral magnesium is the second most abundant intracellular cation and is involved in several important biochemical reactions [8]. It is known that magnesium has antiarrhythmic effect and can influence blood pressure levels by modulating vascular tone. Changes in extracellular magnesium content are able to modify the production and release of nitric oxide (NO), resulting in the alteration of arterial smooth muscle tone by affecting calcium concentrations. Magnesium also participates in glucose metabolism and insulin homeostasis. For these reasons, it has been suggested that magnesium deficiency or changes in its metabolism are related to the pathophysiology of hypertension, atherosclerosis, insulin resistance, and diabetes (Figure 1) [9].

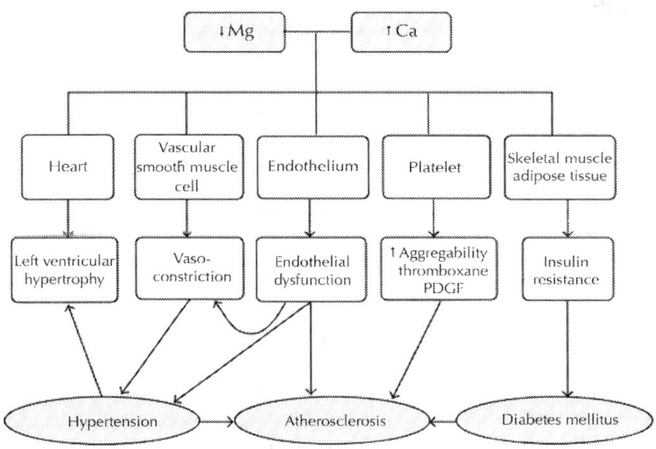


**Figure 1:** Role of magnesium and calcium in the pathophysiology of hypertension, DIABETES MELLITUS, and atherosclerosis.

Increased levels of extracellular magnesium inhibit calcium influx. Conversely, reduced extracellular magnesium activates calcium influx via calcium channels. Low intracellular magnesium concentrations stimulate inositol-trisphosphate-(IP3-) mediated mobilization of intracellular calcium and reduce $Ca^{2+}$-ATPase activity. Thus, calcium efflux and sarcoplasmic reticular calcium reuptake are reduced, LEADING to cytosolic accumulation of calcium and increased intracellular calcium concentration, which is a crucial factor for vasoconstriction. Increased intracellular levels of magnesium result in decreased intracellular free calcium concentration promoting vasodilation [10]. The action of magnesium as a calcium channel blocker may also help to reduce the release of calcium and thus reducing vascular resistance. In addition, magnesium also activates the Na-K ATPase pump that controls the balance of these minerals contributing to the homeostasis of electrolytes in cells [11].

Smaller concentrations of magnesium seem to be associated with reduced serum HDL-cholesterol along with increased LDL-cholesterol and triglycerides levels [9]. Additionally, deficiency of this mineral has been previously related to oxidative stress, proinflammatory state, endothelial dysfunction, platelet aggregation, insulin resistance, and hyperglycemia [12].

High levels of magnesium may increase production of adenosine triphosphate (ATP) and intracellular glucose utilization, since magnesium acts as a cofactor of all reactions involving ATP transfer [13]. Insulin seems to be one of the most important factors that regulate plasma and intracellular magnesium concentrations. It has been suggested that an ATPase-dependent pump is involved in the mechanism by which insulin regulates the erythrocyte magnesium content [14]. On the other hand, intracellular magnesium may play a role in modulating insulin-mediated glucose uptake and vascular tone. Reduced urinary magnesium losses have been implicated in better metabolic control [15]. Low plasma and intracellular magnesium levels may contribute to reducing insulin sensitivity. In fact, suppression of intracellular free magnesium concentrations is known to decrease cellular glucose utilization and thus to promote peripheral insulin resistance as a postreceptor defect [16].

Concerning insulin homeostasis, there is a hypothesis that there is increased secretion of insulin and adrenaline in hypomagnesemia in order to maintain magnesium and cellular cAMP (3′, 5′-cyclic adenosine monophosphate) concentration [17]. Furthermore, the intracellular concentration of magnesium appears to be dependent on the extracellular level, and its influx through calcium channel is voltage dependent. Extracellular magnesium can competitively inhibit calcium channels and determine reduced secretion of insulin. This inhibition does not occur when there is no magnesium in the extracellular space, resulting in higher insulin secretion [18].

Some studies suggest the possible role of intracellular magnesium on the activity as a regulator of the main communication channels of the cell membrane, suggesting that there may be an association between changes in intracellular content of ions induced by supplementation of magnesium and its antihypertensive effects [19].

# MAGNESIUM AND BLOOD PRESSURE

Experimental models of hypertension have been associated with reduced serum and tissue levels of magnesium. In spontaneously hypertensive rats (SHRs), increase of blood pressure arises from the age of young adults, around 12 to 16 weeks of life, being attributed to a genetic component similar to human essential hypertension [20]. In SHR, and also in DOCA-salt model, reduced levels of intracellular magnesium have been noted in smooth muscle cells and cardiomyocytes.

Magnesium supplementation had little antihypertensive effect in adult SHR with well-established hypertension. In fact, the effect of supplementation was only positive in younger animals, when started in the prehypertensive phase, preventing or at least attenuating the development of hypertension [21]. This finding is highly suggestive of a more protective effect of supplemental magnesium, which may prevent or slow the rise in blood pressure at an early stage of hypertension.

In other experimental studies, dietary magnesium deficiency was associated with increased blood pressure in previous normotensive animals, and magnesium supplementation was able to reverse this condition. However, clinical trials of magnesium supplementation in hypertensive patients show divergent results. Some studies demonstrate low serum magnesium levels in hypertensive patients when compared with normotensive subjects, and blood pressure levels reduction after magnesium supplementation [3], although other studies have not confirmed this finding. For this reason, while adequate intake of magnesium through diet is recommended, supplementation of this mineral is not indicated as part of antihypertensive treatment [22, 23].

Experimental, clinical, and epidemiological studies have observed a close inverse relationship between dietary intake or supplementation of magnesium and blood pressure level, indicating the potential role of magnesium deficiency in the pathogenesis of essential hypertension [24], but the mechanism is unclear. The effects of magnesium on the smooth muscle cells growth and inflammation may be important.

A relationship has also been reported between the rennin-angiotensin system, magnesium, and blood pressure. Hypertensive patients with high renin activity have significantly lower serum magnesium levels than normotensive subjects, and plasma renin activity is inversely associated with serum magnesium [25]. Hypertensive patients without blood pressure control may have hypomagnesemia. Hatzistavri and colleagues have shown that magnesium supplementation was associated with slight reduction of 24 h blood pressure levels in patients with mild hypertension [3], which can be evaluated by ambulatory blood pressure monitoring [26]. On the other hand, a study comparing the relationship between serum magnesium, vascular dysfunction, hypertension, and atherosclerosis has not shown enough results to support this association, indicating that low serum magnesium cannot be considered a risk factor for development of these conditions [27].

# MAGNESIUM AND VASCULAR STRUCTURE

Hypertension is also associated with unfavorable changes in elastic properties of large arteries. Some studies have shown the independent prognostic role of arterial stiffness in cardiovascular events in hypertensive patients, which can be assessed by measurements of the pulse wave velocity (PWV) [33–35]. However, there are a few studies showing the influence of magnesium in this condition so far. Van Laecke and colleagues have reported that serum hypomagnesemia associated with hypertension, endothelial dysfunction, dyslipidemia, and inflammation may affect vascular stiffness in patients who underwent kidney transplantation since the low serum magnesium was independently associated with PWV assessed by SphygmoCor [36]. In an experimental study evaluating the structure of the carotid artery in rats, magnesium deficiency was associated with hypertrophic vascular remodeling, which was attenuated by supplementation of this ion. These findings suggest that magnesium deficiency alters the vascular mechanical properties in young animals and may be a mechanism involved in the pathogenesis of hypertension, atherosclerosis, and other cardiovascular diseases [37].

Other possible mechanisms of magnesium action are anti-inflammation, antioxidion, and modulation of cell growth properties. In fact, the production of reactive oxygen species is usually increased in the vasculature of hypertensive patients, and the involvement of magnesium could occur through the reduction of inflammation and oxidative stress [38]. Magnesium has antioxidant properties that could attenuate detrimental effects of oxidative stress on the vasculature, thereby preventing increased vascular tone and contractility [39].

## MAGNESIUM AND VASCULAR FUNCTION

Endothelial dysfunction refers to an imbalance in the endothelial production of mediators that regulate vascular tone, platelet aggregation, coagulation, and fibrinolysis. There is a worsening in the endothelium-dependent relaxation, which can be caused by both loss of NO bioavailability as changes in the production of other endothelium-derived vasoactive substances mainly endothelin-1 and angiotensin II.

The role of magnesium in the endothelial dysfunction has been discussed elsewhere. Indeed, it has been reported that magnesium modifies the vascular tone by regulating endothelium and smooth muscle cell functions along with an important role in the classical pathway of NO release. Experiments in animals have also showed increased production of prostacyclin and NO by magnesium, promoting endothelium-independent and endothelium-dependent vasodilation [40].

The peripheral vascular resistance may be modified by magnesium, also through the regulation of responses to vasoactive agents, particularly angiotensin II, endothelin, and prostacyclin. Animals deficient in magnesium have presented high levels of endothelin-1, whose values have been reduced after supplementation of this mineral [41].

A study that followed more than 90,000 postmenopausal women showed that dietary magnesium intake was inversely associated with plasma concentrations of inflammatory markers such as interleukin-6, C-reactive protein (CRP), and tumor necrosis factor- [7]. This same study emphasized that magnesium intake could improve endothelial dysfunction and inflammation and might play a role in preventing metabolic syndrome.

There are a few studies demonstrating the relationship between magnesium supplementation, endothelial function, arterial stiffness, and carotid intima-media thickness. Some reports point out beneficial effects of magnesium supplementation in improving endothelial function in the brachial artery in patients with coronary artery disease [42], heart failure [43], and DIABETES MELLITUS [44], while others show favorable outcome of magnesium supplementation through improvement of insulin sensitivity [45, 46].

# MAGNESIUM SUPPLEMENTATION

Magnesium can be supplemented in different ways, such as oxide, hydroxide, chelate, sulfate, and citrate. Magnesium sulfate, for example, can be used as anticonvulsant therapy in preeclampsia due to its neuroprotective action and a possible role in regulating vascular tone [47].

Some studies have shown blood pressure lowering after magnesium supplementation. The administration of magnesium oxide (400 mg daily) for eight weeks in patients with hypertension can reduce blood pressure levels, and this reduction has already been detected in office measurements and by ambulatory blood pressure monitoring [29]. A study of 48 subjects has demonstrated that 600 mg of magnesium pidolate per day was able to reduce blood pressure levels in the supplemented patients when compared to the group with no supplementation [3]. This same dosage of supplement was also associated with reduction of serum total cholesterol, LDL-cholesterol, and triglycerides and improvement of insulin resistance.

Haenni and colleagues reported positive effects of magnesium supplementation in order to confirm the relationship between the metabolism of this mineral and alteration of endothelial function by showing increased endothelium-dependent vasodilatation after magnesium infusion [48]. Furthermore, another study showed that chronic magnesium supplementation was able to improve endothelial function in patients with coronary artery disease [42]. Some positive and negative results after magnesium supplementation are shown in Table 1. A meta-analysis evidenced a weak causal correlation between magnesium supplementation and blood pressure reduction, and double-blind placebo controlled trials are needed to determine the effect of magnesium supplementation on cardiovascular outcomes [49].

**Table 1:** Positive and negative results with magnesium supplementation for blood pressure (BP) reduction

| Study population | Mg supplementation | Comparator group | Duration of treatment | Clinical outcome | **Year [Reference]** |
|---|---|---|---|---|---|
| 24 patients with uncomplicated hypertension | 600 mg of magnesium pidolate | 24 age- and sex-matched controls | 12 weeks | Small but significant reductions in mean 24 h systolic and diastolic BP levels | 2009 [18] |
| 35 patients with essential hypertension | magnesium 70.8 mg/d; potassium 217.2 mg/d | 32 patients received lacidipine (4 mg/d) | 4 weeks | Systolic and diastolic BP decreased, and small arterial compliance values increased | 2006 [28] |
| 60 hypertensive patients | 20 mmol/d magnesium oxide | 60 hypertensive patients in a control period, crossover design | 8 weeks | Office, home, and average 24-hour BPs were significantly lower in the magnesium supplementation period | 1998 [29] |
| 15 patients with mild to moderate primary hypertension | 600 mg/day of oxide magnesium | 15 hypertensive patients in a crossover design, receiving placebo | 6 weeks | Significant reduction of systolic, diastolic, and mean BP | 1996 [19] |

| | | | | | |
|---|---|---|---|---|---|
| 698 healthy adults with high-normal diastolic blood pressure | 360 mg of magnesium (diglycine) | 1 g of calcium (carbonate) | 6 months | Neither calcium nor magnesium produced significant changes in BP at 3 and 6 months | 1995 [30] |
| 14 mild to moderate hypertensives | Magnesium pidolate (15 mmol/day) | Placebo | 6 months | Magnesium supplementation does not affect BP at rest and during sympathetic stimulation | 1992 [31] |
| 71 subjects with mild hypertension or a high-normal BP | 15 mmol Mg | Placebo | 6 months | No general effects on the BP | 1991 [32] |

# CONCLUSIONS

Magnesium is a mineral with important functions in the body, and it is important that their levels are adequate. The conflicting results of studies evaluating the effects of magnesium supplements on blood pressure and other cardiovascular outcomes indicate that the action of magnesium in the vascular system is present but not yet established. Certainly, the lack of definitive conclusions due to heterogeneity of study populations with different clinical profiles and severity of illness, lack of standardization of the type of supplement and the dose, and, finally, very short time of treatment, most often between one and three months, are factors that contribute to the difficulty to achieve the primary objectives. Based on recent studies, although we cannot make categorical statements, it appears that magnesium is more involved in the functional vascular changes, and also on local metabolic stability with no influence on the vascular structure. Therefore, further studies are needed to evaluate the risk of magnesium deficiency and the effects to be considered in this mineral supplementation.

Possibly the most important point is to define a more homogeneous study population, considering the same gender and age range, dosage, and type of supplement, as well as a longer period for supplementation. After resolving these concerns, it will be possible to clarify the role of magnesium in the prevention and treatment of cardiovascular diseases.

# REFERENCES

1. B. M. Egan, Y. Zhao, and R. N. Axon, "US trends in prevalence, awareness, treatment, and control of hypertension, 1988–2008," Journal of the American Medical Association, vol. 303, no. 20, pp. 2043–2050, 2010.
2. K. He, K. Liu, M. L. Daviglus et al., "Magnesium intake and incidence of metabolic syndrome among young adults," Circulation, vol. 113, no. 13, pp. 1675–1682, 2006.
3. L. S. Hatzistavri, P. A. Sarafidis, P. I. Georgianos et al., "Oral magnesium supplementation reduces ambulatory blood pressure in patients with mild hypertension," American Journal of Hypertension, vol. 22, no. 10, pp. 1070–1075, 2009.

4. S. Bo and E. Pisu, "Role of dietary magnesium in cardiovascular disease prevention, insulin sensitivity and diabetes," Current Opinion in Lipidology, vol. 19, no. 1, pp. 50–56, 2008.

5. G. Uza, O. Pavel, D. Uza, and R. Valaicu, "Hypomagnesemia in patients with essential arterial hypertension," Magnesium-Bulletin, vol. 9, pp. 177–180, 1987.

6. R. M. Touyz, F. J. Milne, and S. G. Reinach, "Intracellular $Mg^{2+}$, $Ca^{2+}$, $Na^{2+}$ and $K^{+}$ in platelets and erythrocytes of essential hypertension patients: relation to blood pressure," Clinical and Experimental Hypertension A, vol. 14, no. 6, pp. 1189–1209, 1992.

7. S. A. Chacko, Y. Song, L. Nathan et al., "Relations of dietary magnesium intake to biomarkers of inflammation and endothelial dysfunction in an ethnically diverse cohort of postmenopausal women,"Diabetes Care, vol. 33, no. 2, pp. 304–310, 2010.

8. M. Barbagallo, L. J. Dominguez, A. Galioto, A. Pineo, and M. Belvedere, "Oral magnesium supplementation improves vascular function in elderly diabetic patients," Magnesium Research, vol. 23, no. 3, pp. 131–137, 2010.

9. M. Barbagallo, L. J. Dominguez, A. Galioto et al., "Role of magnesium in insulin action, diabetes and cardio-metabolic syndrome X," Molecular Aspects of Medicine, vol. 24, no. 1–3, pp. 39–52, 2003.

10. K. K. Griendling, S. E. Rittenhouse, T. A. Brock, L. S. Ekstein, M. A. Gimbrone Jr., and R. W. Alexander, "Sustained diacylglycerol formation from inositol phospholipids in angiotensin II-stimulated vascular smooth muscle cells," The Journal of Biological Chemistry, vol. 261, no. 13, pp. 5901–5906, 1986.

11. B. Sontia and R. M. Touyz, "Role of magnesium in hypertension," Archives of Biochemistry and Biophysics, vol. 458, no. 1, pp. 33–39, 2007.

12. A. Mazur, J. A. M. Maier, E. Rock, E. Gueux, W. Nowacki, and Y. Rayssiguier, "Magnesium and the inflammatory response: potential physiopathological implications," Archives of Biochemistry and Biophysics, vol. 458, no. 1, pp. 48–56, 2007.

13. M. Barbagallo and L. J. Dominguez, "Magnesium metabolism in type 2 DIABETES MELLITUS, metabolic syndrome and insulin

resistance," Archives of Biochemistry and Biophysics, vol. 458, no. 1, pp. 40–47, 2007.

14. G. Paolisso, S. Sgambato, N. Passariello, et al., "Insulin induces opposite changes in plasma and erythrocyte magnesium concentrations in normal man," Diabetologia, vol. 29, no. 9, pp. 644–647, 1986.
15. C. Schnack, I. Bauer, P. Pregant, P. Hopmeier, and G. Schernthaner, "Hypomagnesaemia in Type 2 (non-insulin-dependent) diabetes mellitus is not corrected by improvement of long-term metabolic control," Diabetologia, vol. 35, no. 1, pp. 77–79, 1992.
16. O. G. Kolterman, R. S. Gray, J. Griffin, et al., "Receptor and postreceptor defects contribute to the insulin resistance in noninsulin-dependent diabetes mellitus," The Journal of Clinical Investigation, vol. 68, no. 4, pp. 957–969, 1981.
17. J. Durlach and V. Durlach, "Speculations on hormonal controls of magnesium homeostasis: a hypothesis," Magnesium, vol. 3, no. 3, pp. 109–131, 1984.
18. L. S. Hadjistavri, P. A. Sarafidis, P. I. Georgianos et al., "Beneficial effects of oral magnesium supplementation on insulin sensitivity and serum lipid profile," Medical Science Monitor, vol. 16, no. 6, pp. CR307–CR312, 2010.
19. A. F. Sanjuliani, V. G. de Abreu Fagundes, and E. A. Francischetti, "Effects of magnesium on blood pressure and intracellular ion levels of Brazilian hypertensive patients," International Journal of Cardiology, vol. 56, no. 2, pp. 177–183, 1996.
20. R. F. Júnior, V. J. D. da Silva, and H. C. Salgado, "Modelos de Hipertensão Arterial," Revista Brasileira de Hipertensão, vol. 8, pp. 19–29, 2001.
21. R. M. Touyz and F. J. Milne, "Magnesium supplementation attenuates, but does not prevent, development of hypertension in spontaneously hypertensive rats," American Journal of Hypertension, vol. 12, no. 8, part 1, pp. 757–765, 1999.
22. A. V. Chobanian, G. L. Bakris, H. R. Black et al., "Seventh report of the Joint National Committee on Prevention, Detection, Evaluation, and Treatment of High Blood Pressure," Hypertension, vol. 42, no. 6, pp. 1206–1252, 2003.

23. G. Mancia, G. de Backer, A. Dominiczak et al., "ESH/ESC 2007 guidelines for the management of arterial hypertension," Revista Espanola de Cardiologia, vol. 60, no. 9, pp. 968.e1–968.e94, 2007.
24. R. M. Touyz, Q. Pu, G. He et al., "Effects of low dietary magnesium intake on development of hypertension in stroke-prone spontaneously hypertensive rats: role of reactive oxygen species," Journal of Hypertension, vol. 20, no. 11, pp. 2221–2232, 2002.
25. L. M. Resnick, J. H. Laragh, J. E. Sealey, and M. H. Alderman, "Divalent cations in essential hypertension: relations between serum ionized calcium, magnesium, and plasma renin activity," The New England Journal of Medicine, vol. 309, no. 15, pp. 888–891, 1983.
26. Y. Kawano, "Role of blood pressure monitoring in non-pharmacological management of hypertension,"Blood Pressure Monitoring, vol. 7, no. 1, pp. 51–54, 2002.
27. A. M. Khan, L. Sullivan, E. McCabe, D. Levy, R. S. Vasan, and T. J. Wang, "Lack of association between serum magnesium and the risks of hypertension and cardiovascular disease," American Heart Journal, vol. 160, no. 4, pp. 715–720, 2010.
28. G. Wu, H. Tian, K. Han, Y. Xi, Y. Yao, and A. Ma, "Potassium magnesium supplementation for four weeks improves small distal artery compliance and reduces blood pressure in patients with essential hypertension," Clinical and Experimental Hypertension, vol. 28, no. 5, pp. 489–497, 2006.
29. Y. Kawano, H. Matsuoka, S. Takishita, and T. Omae, "Effects of magnesium supplementation in hypertensive patients: assessment by office, home, and ambulatory blood pressures," Hypertension, vol. 32, no. 2, pp. 260–265, 1998.
30. M. E. Yamamoto, W. B. Applegate, M. J. Klag et al., "Lack of blood pressure effect with calcium and magnesium supplementation in adults with high-normal blood pressure: results from phase I of the Trials of Hypertension Prevention (TOHP)," Annals of Epidemiology, vol. 5, no. 2, pp. 96–107, 1995.
31. L. A. Ferrara, R. Iannuzzi, A. Castaldo, A. Iannuzzi, A. Dello Russo, and M. Mancini, "Long-term magnesium supplementation

in essential hypertension," Cardiology, vol. 81, no. 1, pp. 25–33, 1992.

32. L. Lind, H. Lithell, T. Pollare, and S. Ljunghall, "Blood pressure response during long-term treatment with magnesium is dependent on magnesium status. A double-blind, placebo-controlled study in essential hypertension and in subjects with high-normal blood pressure," American Journal of Hypertension, vol. 4, no. 8, pp. 674–679, 1991.
33. S. Laurent, P. Boutouyrie, R. Asmar et al., "Aortic stiffness is an independent predictor of all-cause and cardiovascular mortality in hypertensive patients," Hypertension, vol. 37, no. 5, pp. 1236–1241, 2001.
34. A. Benetos, C. Adamopoulos, J. M. Bureau et al., "Determinants of accelerated progression of arterial stiffness in normotensive subjects and in treated hypertensive subjects over a 6-year period,"Circulation, vol. 105, no. 10, pp. 1202–1207, 2002.
35. A. Adji, M. F. O'Rourke, and M. Namasivayam, "Arterial stiffness, its assessment, prognostic value, and implications for treatment," American Journal of Hypertension, vol. 24, no. 1, pp. 5–17, 2011.
36. S. van Laecke, C. Maréchal, F. Verbeke et al., "The relation between hypomagnesaemia and vascular stiffness in renal transplant recipients," Nephrology Dialysis Transplantation, vol. 26, no. 7, pp. 2362–2369, 2011.
37. P. Laurant, D. Hayoz, H. Brunner, and A. Berthelot, "Dietary magnesium intake can affect mechanical properties of rat carotid artery," British Journal of Nutrition, vol. 84, no. 5, pp. 757–764, 2000.
38. R. M. Touyz and E. L. Schiffrin, "Reactive oxygen species in vascular biology: implications in hypertension," Histochemistry and Cell Biology, vol. 122, no. 4, pp. 339–352, 2004.
39. P. Laurant and R. M. Touyz, "Physiological and pathophysiological role of magnesium in the cardiovascular system: implications in hypertension," Journal of Hypertension, vol. 18, no. 9, pp. 1177–1191, 2000.
40. C. A. Northcott and S. W. Watts, "Low $[Mg^{2+}]e$ enhances arterial spontaneous tone via phosphatidylinositol 3-kinase in DOCA-salt hypertension," Hypertension, vol. 43, no. 1, pp. 125–129, 2004.

41. P. Laurant and A. Berthelot, "Endothelin-1-induced contraction in isolated aortae from normotensive and DOCA-salt hypertensive rats: effect of magnesium," British Journal of Pharmacology, vol. 119, no. 7, pp. 1367–1374, 1996.
42. M. Shechter, M. Sharir, M. J. Labrador et al., "Oral magnesium therapy improves endothelial function in patients with coronary artery disease," Circulation, vol. 102, no. 19, pp. 2353–2358, 2000.
43. D. Almoznino-Sarafian, G. Sarafian, S. Berman et al., "Magnesium administration may improve heart rate variability in patients with heart failure," Nutrition, Metabolism and Cardiovascular Diseases, vol. 19, no. 9, pp. 641–645, 2009.
44. F. Guerrero-Romero and M. Rodríguez-Morán, "The effect of lowering blood pressure by magnesium supplementation in diabetic hypertensive adults with low serum magnesium levels: a randomized, double-blind, placebo-controlled clinical trial," Journal of Human Hypertension, vol. 23, no. 4, pp. 245–251, 2009.
45. F. Guerrero-Romero, H. E. Tamez-Perez, G. González-González et al., "Oral Magnesium supplementation improves insulin sensitivity in non-diabetic subjects with insulin resistance: a double-blind placebo-controlled randomized trial," Diabetes and Metabolism, vol. 30, no. 3, pp. 253–258, 2004.
46. S. Lee, H. K. Park, S. P. Son, C. W. Lee, I. J. Kim, and H. J. Kim, "Effects of oral magnesium supplementation on insulin sensitivity and blood pressure in normo-magnesemic nondiabetic overweight Korean adults," Nutrition, Metabolism and Cardiovascular Diseases, vol. 19, no. 11, pp. 781–788, 2009.
47. R. M. Touyz, "Role of magnesium in the pathogenesis of hypertension," Molecular Aspects of Medicine, vol. 24, no. 1–3, pp. 107–136, 2003.
48. A. Haenni, K. Johansson, L. Lind, and H. Lithell, "Magnesium infusion improves endothelium-dependent vasodilation in the human forearm," American Journal of Hypertension, vol. 15, no. 1, part 1, pp. 10–15, 2002.
49. H. O. Dickinson, D. J. Nicolson, F. Campbell et al., "Magnesium supplementation for the management of essential hypertension

in adults," Cochrane Database of Systematic Reviews, vol. 3, Article ID CD004640, 2006.

Chapter 5

# Structural and Functional Alterations in Photosynthetic Apparatus of Plants under Cadmium Stress

Pooja Parmar, Nilima Kumari, and Vinay Sharma

Department of Bioscience and Biotechnology, Banasthali University, P.O. Banasthali Vidyapith, Rajasthan -304022, India

## ABSTRACT

Cadmium is a potentially toxic heavy metal that hampers plant productivity by interfering with their photochemistry. Cd causes disturbances in a range of physiological processes of plants such as photosynthesis, water relations, ion metabolism and mineral uptake. Cd pronouncedly affects photosynthesis by alteration of its vital machinery in all aspects. Photosynthesis is a well organised and sequential process fundamental to all green plants and microorganisms which involves various components, including photosynthetic pigments and photosystems, the electron transport system and $CO_2$ reduction pathways. Any damage at any level caused by Cd, critically affects overall photosynthetic capacity. Present review focuses on key effects of Cd on photosynthetic apparatus including chloroplast structure, photosynthetic pigments, Chl-protein complexes and photosystems resulting in overall decrease in efficiency of carbon assimilation pathway.

# REVIEW

## Introduction

The unprecedented increase in heavy metal pollution has become a matter of major concern over the globe (Jamali et al. 2007). Cadmium (Cd) stands 7th out of the 20 toxins and has no known biological function except in marine diatoms (Morel 2008). Cd is used and traded internationally as a metal and as a chemical compound throughout Asia, America, Europe, Australia and Africa (UNEP 2010). Indeed, Cd concentration is progressively increasing at an alarming rate (7 to 43 percent over the period of 100 years) in several European countries such as Austria, Denmark, Finland, Greece, Ireland and the United Kingdom (UNEP 2010). It has been estimated that major source of Cd release into the air are the production of nonferrous metals followed by iron and steel production, fossil fuel combustion, cement production and waste incineration (Pacyna and Pacyna2001). Cd is constantly added and gets accumulated to the plough layer of soil through various natural and anthropogenic activities such as volcanic eruptions, mining, smelting, mismanagement of industrial waste and use of phosphate fertilizers (Grant 2011) and its addition to the arable land is a widely recognised problem. Cd is potentially toxic to all organisms including plants, animals and humans as well. Cd exposure, for instance, is associated with cancers of the prostate, lungs and testes, kidney tubule damages, rhinitis, emphysema, osteomalacia and bone fractures in humans (Nawrot et al. 2006). In plants, it results in many toxic symptoms such as inhibition of growth and photosynthesis, activation or inhibition of enzymes, disturbances in plant-water relations and ion metabolism, and formation of free radicals (Valentoviova et al. 2010).

Phytotoxicity induced by Cd has been well established and comprehensively studied (Wahid et al.2009). Cd is taken up by roots through plasma membrane transporters such as ZIP (ZRT-IRT like protein;

Zinc regulated transporter, Iron-regulated transporter) and NRAMP (natural resistance associated macrophage protein) in competition to the essential nutrients of plants (Kim et al. 2002) and consequently it is translocated to shoots thereby leading to growth diminution which in due part emanates from disturbed photosynthesis (Bazzaz et al. 1974). Figure 1 illustrates the effects of Cd as a potent inhibitor of photosynthesis. Photosynthesis inhibition may be attributed to diminished chlorophyll biosynthesis (Shukla et al. 2008), interrupted $O_2$ - evolving reactions of PSII and altered electron flow around PSI and PSII (Mallick and Mohn 2003). Cd hampers Calvin cycle by slowing down activity of various enzymes hence resulting in decreased photosynthesis (Ying et al. 2010). Cd has also been known to show inhibitory effect on various enzymes such as ribulose-1,5-biphosphate carboxylase oxygenase (Mobin and Khan 2007), phosphoenolpyruvate carboxylase (Latif 2008), aldolase (Sheoran et al. 1990), fructose-6-phosphate kinase (Malik et al.1992), fructose-1,6-bisphosphatase (Sheoran et al. 1990), $NADP^+$-glyceraldehyde-3-phosphate dehydrogenase (Sheoran et al. 1990) and carbonic anhydrase (Mobin and Khan 2007).

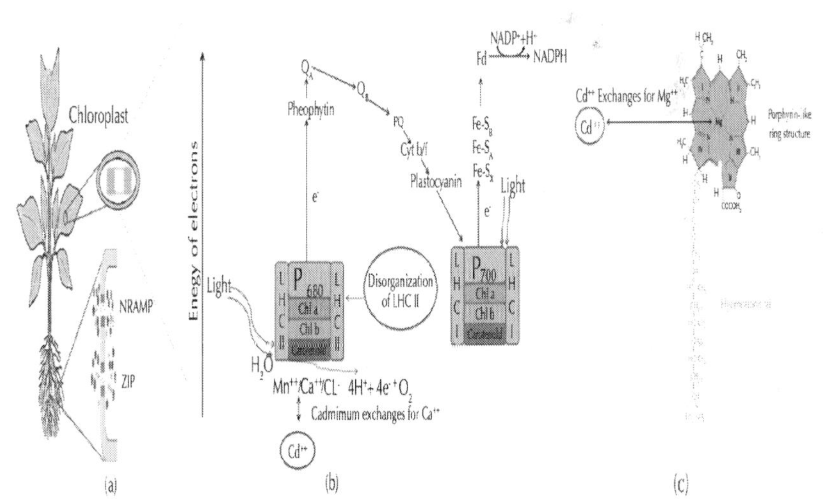


**Figure 1:** An overview of effects of Cd exposure to plants at different levels in photosynthetic machinery. (a) Cd uptake in cells through plasma membrane transporters. (b) Alteration in organisation of oxygen evolving and light harvesting complexes, Cd also binds with $Q_B$ pocket thus slows down electron flow from $Q_A$ to $Q_B$. (c) Incorporation of Cd in chlorophyll molecule.

Stomatal closure due to entry of Cd into the guard cells in competition to $Ca^{+2}$ (Perfus-Barbeoch et al. 2002) and reduction in stomata count per unit area are also characteristic symptoms of Cd stress resulting in lesser conductance to $CO_2$ (Pietrini et al. 2010) which consequently leads to overall inhibition of photosynthesis.

The present review is an attempt to develop an orchestrated understanding of the mechanisms involved in altering and damaging various components of photosynthetic machinery by Cd thereby leading to effective loss in the anabolic reactions of plants.

## Photosynthetic Machinery under Cd Stress

### *Chloroplast Structure*

Cd convincingly resulted in marked distortion of chloroplast ultrastructure leading to disturbed shape and inflated thylakoids (Najeeb et al. 2011). Chloroplast structure disturbance has been partly manifested by a notable decrease in chloroplast number and size, grana stacking, starch grain content and accumulation of plastoglobuli observed in various plants such as *Picris divarticata* (75 µM, 14 days after treatment (DAT)), *Hordeum vulgare* (5 µM, 15 DAT) and *Brassica*(Ying et al. 2010; Wang et al. 2011; Elhiti et al. 2012). Further, plants show differential aggregation of grana in young and older leaves. For instance in willow, older leaves showed swollen but organised thylakoids whereas young leaves appear to be more dense structured accompanied by tannin precipitation. Reed plant chloroplasts displayed a disturbed shape, wavy appearance of grana and stroma thylakoids and swollen intra thylakoidal space owing to lipid peroxidation, a consequence of increased lipid accumulation in thylakoids (Hakmaouia et al. 2007).

Disruption in chloroplast structure is also ensued due to increased peroxidation of membrane fatty acid and lipid contents resulting from enhanced lipooxygenase (LOX) activity (Remans 2010). LOX mediates polyunsaturated fatty acid oxidation including chloroplast membrane lipids such as monogalactosyldiacyl-glycerol (MGDG), digalactosyldiacyl-glycerol (DGDG) and phosphatidyl glycerol (PG) hence resulting in production of free radicals. LOX activity has been positively correlated with increased lipid peroxidation in plants such as

*Arabidopsis*, Barley, Lupine and*Phaseolus* under Cd stress (Maksymiec and Krupa 2006; Tamas et al. 2009). A significant decrease has also been reported in the content of polaracyl lipids especially MGDG, DGDG and PG in tomato chloroplasts membranes (Djebali et al. 2005) which is considered to be indispensable for maintenance of membrane integrity.

Grana disorganization can be attributed to reduced MGDG level, as well as the decrease in 16:1 trans fatty acid content in MGDG and PG. In *Brassica napus* (50 μM, 15 DAT) leaves, remarkable decrease upto 80–84 % was observed in DGDG and MGDG respectively (Nouairi et al. 2005), which may possibly be a reason in disintegrated grana.

## Cd Induced Pigment Changes

Among the photosynthetic pigments enormous studies have been conducted till date on reduction in chlorophyll and carotenoids in plants exposed to Cd stress. Chlorophyll destruction in older leaves and its biosynthesis inhibition in newer ones have been known to be prime cause in leaf chlorosis in plants growing in Cd treated soils (Xue et al. 2013). Inhibition of chlorophyll biosynthesis enzymes and activation of its enzymatic degradation plays crucial role in net loss in chlorophyll content (Somashekaraiah 1992).

Aminolevulinate (ALA) is a crucial compound in chlorophyll biosynthesis and its synthesis is the rate-limiting and regulatory step. Cd inhibits ALA synthesis at the site of availability of glutamate for ALA synthesis and interferes by interacting with SH group of enzymes, δ-aminolevulinic acid dehydratase (Mysliwa-Kurdziel and Strzalka 2002) and porphobilinogen deaminase, (Skrebsky et al. 2008) leading to the accumulation of chlorophyll biosynthesis intermediates like ALA and porphyrins. In fact ALA accumulation is considered to be a reason for generation of reactive oxygen species which alters redox status of plants and thus disturbing plant homeostasis as reported in *Soybean* (0-100 μM, 10 DAT) and *Cucumis* (0-1000 μM, 10 DAT) (Noriega et al. 2007; Goncalves et al. 2009). Additionally Cd reacts with protochlorophillide reductase, which causes photoreduction of protochlorophillide into chlorophyllide thus diminishing the raw material for chlorophyll synthesis (Stobart et al. 1985).

Cd also decreases uptake of nutrients such as Mn, Fe and Mg, hence a comparatively higher amount of cellular Cd interferes with $Mg^{2+}$ insertion into protoporphyrinogen or may cause Chl destruction as consequence of $Mg^{2+}$ substitution in both Chl a and b (Gillet et al. 2006).

Carotenoid content in plants exposed to Cd do not exhibit a set pattern and may either increase or decrease. The increase has been observed in many cases as in *Cucumis sativus* L. (Burzynski et al. 2007) and *Zea mays* L. (100 μM, 10 DAT) (Chaneva et al. 2010). On the contrary decrease was also observed in a few cases e.g. *Pisum sativum* (7 mg/kg, 20 DAT) (Hattab et al. 2009). Other leaf pigments including neoxanthin, lutein, violaxanthin were found to decrease in *Lycopersicon esculentum* and *Spinacea oleracea* plants (López-Millán et al. 2009; Fagioni et al. 2009).

In lower organisms, Cd exposure caused a significant drop in the amounts of phycobiliprotein viz. allophycocyanin, phycocyanin, and phycoerythrin e.g. *Chlamydomonas* (50 μM, for 24 hrs),*Gracilaria* (300 μM, 16 DAT) and *Hypnea musciformis* (300μM, 7DAT) that led to decrease in photosynthetic efficiency (Perrault et al. 2011; Santos et al. 2012; Bouzon et al. 2012).

## Cd Induced Changes in Chlorophyll-Proteins Complexes

Chl-proteins can be described as Chl a and Chl a/b multicofactor proteins for both photosystems (PS) bound to chlorophylls and carotenoids (Fromme et al. 2001). Cd effects on both the PS as well as degree of damage vary in the plant species even among cultivars and populations, depending on genotypic and ecotypic differences (Prasad 1995).

## PSII Core Complex

Immunoblotting of Chl-protein complexes did not depict any changes in the level of polypeptides of PSII complexes comprising of CP 47, CP 43, D1 and D2 under Cd stress as demonstrated in rice (75 μM Cd, 28 DAT) and spinach (100 μM Cd, 30 DAT) (Pagliano et al. 2006; Fagioni et al. 2009). The same pattern was also observed in lower organisms i.e. *Chlamydomonas reinhardtii* (50 μM, for 24 hrs) too (Perreault et al. 2011).

Cd toxicity may be attributed to both acceptor and donor side of PSII thus preventing photoactivation (Sigfridsson et al. 2004). On the donor side due to high affinity, Cd exchanges with $Ca^{++}$ in $Mn^{++}$/$Ca^{++}$ cofactor present in oxygen evolving complex (Faller et al. 2005; Pagliano et al. 2006); the exchange leads to reduced kinetics of Hill reaction. On acceptor side Cd decreased the rate of electron transfer from $Q_A$ to $Q_B$ due to interaction with nonheme Fe and conformational modification of $Q_B$ pocket (Geiken et al. 1998).

Further decrease in lipid content in chloroplasts specifically MGDG and DGDG (Nouairi et al. 2005), considered to be indispensable for PSII activity, causes structural disorganisation of PSII supramolecular structure (Quartacci et al. 2000).

## Light Harvesting Complex (LHC) II

LHCII is the principle light harvesting pigment-protein complex of PSII which absorbs light energy and transfers it to the reaction centre. The native form of LHCII is a trimer composed of three Lhcb proteins: Lhcb1, Lhcb2 and Lhcb3 (Lucinski and Jackowski 2006). These LHCII aggregates play dynamic role in triggering the thermal dissipation of extra energy for efficient excitation quenching and display photoprotective role in case of overexcitation of reaction centre and antenna (Barros et al. 2009). Cd exposure results in dissipation of total mass of Lhcb1 and Lhcb2 and accounts for disorganization of trimer-forming monomers resulting in diminished LHCII aggregation complexes. This was indicated by infrared studies on *Secale cereale* exposed to Cd (50 μM, 7 DAT) where aggregate/trimeric ratio remained 73% of the control (Janik et al. 2010). Cd toxicity resulted in constrained dissipation of excitation energy which may have been induced by alterations in the quenching centre formation or inhibition of vibrational transfer of thermal energy between pigments and the protein skeleton (Gruszecki et al. 2009). In *Spinacia oleracea* L. Lhcb1.1 isomers of Lhcb1 were highly affected even in small exposure to stress (75 μM, 5 DAT) whereas others i.e. Lhcb2 and Lhcb3 were less affected (Fagioni et al. 2009). Differential level of expression in Lhcb2 was observed in case of two ecotypes of *Sedum* (hyperaccumulating and non-hyperaccumulating) which suggested temporal regulation of gene expression. Upon 24 hrs of Cd (2μ M) treatment non-hyperaccumulating ecotype exhibited higher expression level than hyperaccumulating, followed by a reversal of the situation after 8 days (Zhang et al. 2011).

Proteomic studies on *Oryza sativa* L. (7.5-75μM, 24 DAT) suggested contrasting results where LHCII content is not adversely affected suggesting that antenna complexes of PSII are less affected (Pagliano et al. 2006).

Lipid profiling of chloroplast is conducive to suggest that decrease in 16:1 trans fatty acid content in MGDG and PG, diminished LHCII oligomerization due to its specific binding in sn-2 position in the chloroplastic PG (Vassilev et al. 2004). Cadmium due to its high affinity gets substituted in pigment protein complexes causing conformational changes (Küpper et al. 2002) leading to incorrect binding of chlorophyll molecule to the protein matrix.

## PSI Core Complexes

In some plants exposed to cadmium stress PSI instead of PSII is the prime site of damage. Previous studies suggested that Cd induced iron deficiency in cell organelles is possibly a reason for greater damage to PSI (Siedlecka and Baszyński 1993; Timperio et al. 2007). Prolonged deficiency of iron resulted in generation of reactive oxygen species in thylakoids which principally destroys iron-sulphur centres (PSI) and Lhca antennae (Michel and Pistorius 2004). In fact, observations suggesting damage to PSI have been reported in *Cucumis sativus* L. (10 μM, 35 DAT) (Sárvári 2005; Sárvári et al. 2008) and wheat (Atal et al. 1991). However in *Pisum sativum*extended stress treatment of Cd (0–10 mM, 12 DAT) led to equal damage to both PSI and PSII (Chugh and Sawhney 1999).

Proteomic and expression studies conducted on basal leaves of *Spinacia oleracea* L. (100 μM, 0–15 DAT) revealed presence of modified amino acids in polypeptide chains of PsaA/PsaB proteins corroborating the accumulation of incomplete monomeric units leading to disruption of PSI supercomplexes, reaction centre I and LHCI (Fagioni et al. 2009).

## Cadmium Induced Changes in Photosynthetic Yield

The main effect of Cd, studied to date is hampering the photochemical activity of both PSI and PSII. The reports regarding this have always been contradictory in deciding the principal site of damage i.e. PSI or PSII. As observed in peas, Cd affects both the photosystems over a long period of stress (Chugh and Sawhney 1999). However, during initial stages Cd had more pronounced effect on the activity of PSII as observed in *Thalspi caerulescence* (Kupper et al.2007) stating higher sensitivity of PSII to Cd toxicity (Wang et al. 2013).

The chlorophyll fluorescence induction parameters represent the use of non-invasive tool to understand the photosynthetic performance in vivo and to assess effects of stress on plants photochemistry (Baker et al. 2008). Table 1 depicts effects of Cd on some of the photosynthetic parameters [Chl (chlorophyll content), Fv/Fm (maximum quantum efficiency of PSII) and $P_N$ (Net photosynthetic rate)] which reflect lesions in plants photosynthetic yield as a consequence of damage to the photosystems and pigments (Wang et al. 2009). It is evident from the Table 1 that Cd decreased chlorophyll content, $F_V/F_M$ and $P_N$ indicating impeded photosynthesis. However, PSII disruption as a consequence of Cd toxicity is reported to depend on the irradiance conditions. During high light intensity, direct damage to the PSII reaction centre occurs instead; this was termed as the 'sun reaction'. On the contrary, LHCII disruption due to exchange of $Mg^{2+}$ with Cd in chlorophyll pigment is the prime cause of diminished PSII activity during dark phase and is referred as 'shade reaction' (Kupper et al. 2007).

**Table 1:** Effects of Cd on photosynthetic parameters[1]

| Plant species | Cd concentration | % inhibition | | | | Reference |
|---|---|---|---|---|---|---|
| | | **Chl a** | **Chl b** | **Fv/ Fm** | **Pn** | |
| *Pisum sativum* | 6 mM, 7 DAT | 31.00 | 32.30 | --* | 79.90 | ; Januškaitienė, 2012 |
| *Picris divarticata* | 75 µM, 10 DAT | 10.84 | 1.29 | --* | 44.20 | Ying et al. 2010 |

| | | | | | | |
|---|---|---|---|---|---|---|
| *Zea mays* | 20 µM, 7 DAT | 36.71 | 37.93 | 3.76 | 13.00 | Wang et al. 2009 |
| *Ricinus communis* | 50 µM, 12 DAT | 27.30 | 16.70 | 5.00 | 47.00 | Liu et al. 2011 |
| *Lycopersicon esculentum* | 100 µM, 12 DAT | 35.00 | 23.00 | 2.47 | 73.00 | ;López-Millán et al. 2009 |
| *Cucumis sativus* | 50 µM, 48 hrs | 13.89 | 24.44 | 1.57 | 73.00 | ; Burzynski & Zurek, 2007 |
| *Phragmites australis* | 100 µM, 21 DAT | 52.30 | 73.00 | 1.28 | 40.88 | Pietrini et al. 2003 |

* Values are not available.

1Effective concentration of Cd causing reduction in chlorophyll pigments viz. Chl a and Chl b and non-invasive parameters viz. photosynthetic yield (Fv/Fm) and net CO2 assimilation rate (Pn).

Parmar et al.

Parmar et al. Botanical Studies 2013 54:45 doi:10.1186/1999-3110-54-45

Comparatively less information is known to us in case of PSI but still data exists which shows higher sensitivity to PSI photochemistry as in *Pisum sativum* plants (Wodala et al. 2012).

# CONCLUSIONS

In conclusion, Cd affects photosynthesis either directly or indirectly thus decreasing the crop yield. We reviewed its inhibitory effect on pigments, lipids, photosystems proteins and chloroplasts. Summing up all we investigated net loss in photosynthesis. It can be said that much has been known about Cd toxicity to plants but numerous mechanisms remains debatable about its interaction with photosynthetic proteins i.e. D1 and D2 and oxygen evolving complexes. In particular, we should extend our knowledge towards PSI measurements to get an intricate knowledge on effect of Cd on photosynthesis. Strategies must be evolved on understanding the mechanism of Cd hyperaccumulation to uphold various phytoremediation strategies.

# AUTHORS' CONTRIBUTIONS

PP and NK surveyed the literature and prepared the draft. VS provided guidelines for the review, modified and wrote the final version. All authors read and approved the final manuscript.

# REFERENCES


1. Atal N, Saradhi PP, Mohanty P (1991) Inhibition of the chloroplast photochemical reactions by treatment of wheat seedlings with low concentrations of cadmium: Analysis of electron transport activities and changes in fluorescence yield. Plant Cell Physiol 32:943-951
2. Baker NR (2008) Chlorophyll fluorescence: a probe of photosynthesis in vivo. Annu Rev Plant Biol 59:89-113
3. Barros T, Royant A, Standfuss J, Dreuw A, Kühlbrandt W (2009) Crystal structure of plant light-harvesting complex shows the active, energy-transmitting state. EMBO J 28:298-306.
4. Bazzaz MB, Govindjee (1974) Effects of cadmium nitrate on spectral characteristics and light reactions of chloroplasts. Environ Lett 6:1-2.
5. Bouzon ZL, Ferreira EC, dos Santos R, Scherner F, Horta PA, Maraschin M, Schmidt ÉC (2012) Influences of cadmium on fine structure and metabolism of *Hypnea musciformis*(Rhodophyta, Gigartinales) cultivated in vitro. Protoplasma 249:637-650
6. Burzynski M, Zurek A (2007) Effects of copper and cadmium on photosynthesis in cucumber cotyledons. Photosynthetica 45:239-244
7. Chaneva G, Parvanova P, Tzvetkova N, Uzunova A (2010) Photosynthetic response of maize plants against cadmium and paraquat impact. Water Air Soil Pollut 208:287-293
8. Chugh LK, Sawhney SK (1999) Photosynthetic activities of *Pisum sativum* seedlings grown in presence of cadmium. Plant Physiol Biochem 37:297-303
9. Djebali W, Zarrouk M, Brouquisse R, Kahoui ES, Limam F, Ghorbel MH, Chaïbi W (2005) Ultrastructure and lipid alterations induced

by cadmium in tomato (*Lycopersicon esculentum*) chloroplast membranes. Plant Biol 7:358-368

10. Elhiti M, Yang C, Chan A, Durnin DC, Belmonte M, Ayele BT, Tahir M, Stasolla C (2012) Altered seed oil and glucosinolate levels in transgenic plants over-expressing the *Brassica napus shoot meristem less* gene. J Exp Bot 63:4447-4461
11. Fagioni M, D'Amici GM, Timperio AM, Zolla L (2009) Proteomic analysis of multiprotein complexes in the thylakoids membrane upon cadmium treatment. J Proteome Res 8:310-326
12. Faller P, Kienzler K, Krieger-Liszkay A (2005) Mechanism of $Cd^{2+}$ toxicity: $Cd^{2+}$ inhibits photoactivation of photosystem II by competitive binding to the essential $Ca^{2+}$ site. Biochim Biophys Acta 1706:158-164
13. Fromme P, Jordan P, Krauss N (2001) Structure of photosystem. Biochim Biophys Acta 1507:5-31
14. Geiken B, Masojidek J, Rizzuto M, Pompili ML, Giardi MT (1998) Incorporation of S-35 methionine in higher plants reveals that stimulation of the D1 reaction centre II protein turnover accompanies tolerance to heavy metal stress. Plant Cell Environ 21:1265-1273
15. Gillet S, Decottignies P, Chardonnet S, Le Maréchal P (2006) Cadmium response and redoxin targets in *Chlamydomonas reinhardtii*: a proteomic approach. Photosynth Res 89:201-211
16. Goncalves JF, Nicoloso FT, Becker AG, Pereira LB, Tabaldi LA, Cargnelutti D, dePelegrin CMG, Dressler VL, da Rocha JBT, Schetinger MRC (2009) Photosynthetic pigments content, δ-aminolevulinic acid dehydratase and acid phosphatase activities and mineral nutrients concentration in cadmium-exposed *Cucumis sativus* L. Biologia 64:310-318.
17. Grant CA (2011) Influence of phosphate fertilizer on cadmium in agricultural soils and crops. Agric Agri Food Canada 54:143-155
18. Gruszecki W, Janik E, Luchowski R, Kernen P, Grudziński W, Gryczyński I (2009) Supramolecular organization of the main photosynthetic antenna complex LHCII: a monomolecular study. Langmuir 25:9384-9391
19. Hakmaouia A, Atera M, Bókab K, Barónc M (2007) Copper and cadmium tolerance, uptake and effect on chloroplast ultrastructure studies on *Salix purpurea* and *Phragmites australis*.

Z Naturforsch C 62:417-426

20. Hattab S, Dridi B, Chouba L, Kheder MB, Bousetta H (2009) Photosynthesis and growth responses of pea *Pisum sativum* L. under heavy metals stress. J Environ Sci 21:1552-1556
21. Jamali MK, Kazi TG, Arain MB, Afridi HI, Jalbani N, Memon AR (2007) Heavy metal contents of vegetables grown in soil, irrigated with mixtures of wastewater and sewage sludge in Pakistan, using ultrasonic-assisted pseudo-digestion. J Agron Crop Sci 193:218-228
22. Janik E, Maksymiec W, Mazur R, Garstka M, Gruszecki WI (2010) Structural and functional modifications of the major light-harvesting complex II in cadmium or copper-treated *Secale cereale*. Plant Cell Physiol 51(8):1330-1340
23. Januškaitienė I (2012) the effect of cadmium on several photosynthetic parameters of pea (*Pisum sativum* L.) at two growth stages. Agriculture 1(99):71-76
24. Kim YY, Yang YY, Lee Y (2002) Pb and Cd uptake in rice roots. Physiol Plant 116:368-372
25. Küpper H, Šetlík I, Spiller M, Küpper FC, Prášil O (2002) Heavy metal-induced inhibition of photosynthesis: targets of in vivo heavy metal chlorophyll formation. J Phycol 38:429-441
26. Küpper H, Parameswaran A, Leitenmaier B, Trtílek M, Šetlík I (2007) Cadmium-induced inhibition of photosynthesis and long-term acclimation to cadmium stress in the hyperaccumulator *Thlaspi caerulescens*. New Phytol 175:655-674
27. Latif AA (2008) Cadmium induced changes in pigment content, ion uptake, proline content and phosphoenolpyruvate carboxylase activity in *Triticum aestivum* seedlings. Aust J Bas App Sci 2:57-62
28. Liu C, Guo J, Cui Y, Lu T, Zhang X, Shi G (2011) Effects of cadmium and salicylic acid on growth, spectral reflectance and photosynthesis of castor bean seedlings. Plant Soil 344:131-141
29. López-Millán AF, Sagardoy R, Solanas M, Abadía A, Abadía J (2009) Cadmium toxicity in tomato (*Lycopersicon esculentum*) plants grown in hydroponics. Environ Exp Botany 65:376-385
30. Luciński R, Jackowski G (2006) The structure, functions and degradation of pigment-binding proteins of photosystem II. Acta

Biochim Pol 53:693-708

31. Maksymiec W, Krupa Z (2006) The effects of short-term exposition to Cd, excess Cu ions and jasmonate on oxidative stress appearing in *Arabidopsis thaliana*. Environ Exp Bot 57:187-194
32. Malik D, Sheoran IS, Singh R (1992) Carbon metabolism in leaves of cadmium treated wheat seedlings. Plant Physiol Biochem 30:223-229
33. Mallick N, Mohn FH (2003) Use of chlorophyll fluorescence in metal-stress research: a case study with the green microalga *Scenedesmus*. Ecotoxicol Environ Saf 55:64-69.
34. Michel KP, Pistorius EK (2004) Adaptation of the photosynthetic electron transport chain in cyanobacteria to iron deficiency: the function of IdiA and IsiA. Physiol Plant 120:36-50.
35. Mobin M, Khan NA (2007) Photosynthetic activity, pigment composition and antioxidative response of two mustard (*Brassica juncea*) cultivars differing in photosynthetic capacity subjected to cadmium stress. J Plant Physiol 164:601-610 |
36. Morel FMM (2008) The co-evolution of phytoplankton and trace element cycles in the oceans. Geobiology 6:318-324
37. Mysliwa-Kurdziel B, Strzalka K (2002) Influence of metals on the biosynthesis of photosynthetic pigments. In: Prasad MNV, Strzalka K (eds) Physiology and biochemistry of metal toxicity and tolerance in plants, Netherlands: Springer. pp 201-228
38. Najeeb U, Jilanic G, Alia S, Sarward M, Xua L, Zhoua W (2011) Insights into cadmium induced physiological and ultra-structural disorders in *Juncus effusus* L. and its remediation through exogenous citric acid. J Hazard Mater 186:565-574
39. Nawrot TS, Hecke VE, Thijs L, Richart T, Kuznestsova T, Jin Y, Vangronsveld J, Roels HA, Staessen JA (2006) Environmental exposure to cadmium and risk of cancer: A prospective population-based study. Lancet Oncol 7:119-126
40. Noriega GO, Balestrasse KB, Batlle A, Tomaro ML (2007) Cadmium induced oxidative stress in soybean plants also by the accumulation of δ-aminolevulinic acid. Biometals 20:841-851.
41. Nouairi I, Ammar WB, Youssef NB, Daoud DBM, Ghorbal MH, Zarrouk M (2005) Comparative study of cadmium effects on membrane lipid composition of *Brassica juncea*and *Brassica napus* leaves. Plant Sci 170:511-519

42. Pacyna JM, Pacyna EG (2001) An assessment of global and regional emissions of trace metals to the atmosphere sources worldwide. Environ Rev 9:269-298

43. Pagliano C, Raviolo M, Vecchia FD, Gabbrielli R, Gonnelli C, Rascio N, Barbato R, Rocca NL (2006) Evidence for PSII donor-side damage and photoinhibition induced by cadmium treatment on rice (*Oryza sativa* L.). J Photochem Photobiol B Biol 84:70-78

44. Perfus-Barbeoch L, Leonhardt N, Vavasseur A, Forestier C (2002) Heavy metal toxicity: cadmium permeates through calcium channels and disturbs the plant water status. Plant J 32:539-548

45. Perreault F, Dionne J, Didur O, Juneau P, Popovic R (2011) Effect of cadmium on photosystem II activity in *Chlamydomonas reinhardtii*: alteration of OJIP fluorescence transients indicating the change of apparent activation energies within photosystem II. Photosynth Res 107:151-157

46. Pietrini F, Iannelli MA, Pasqualini S, Massacci A (2003) Interaction of cadmium with glutathione and photosynthesis in developing leaves and chloroplasts of *Phragmites australis* (Cav.) Trin. ex steudel. Plant Physiol 133(2):829-837

47. Pietrini F, Zacchini M, Iori V, Pietrosanti L, Ferretti M, Massacci A (2010) spatial distribution of cadmium in leaves and on photosynthesis: examples of different strategies in willow and poplar clones. Plant Biol 12:355-363

48. Prasad MNV (1995) Cadmium toxicity and tolerance in vascular plants. Environ Exp Bot 35:525-545

49. Quartacci MF, Pinzino C, Sgherri CLM, Vecchia FD, NavariIzzo F (2000) Growth in excess copper induces changes in the lipid composition and fluidity of PSII-enriched membranes in wheat. Physiol Plant 108:87-93

50. Remans T, Opdenakker K, Smeets K, Mathijsen D, Vangronsveld J, Cuypers A (2010) Metal-specific and NADPH oxidase dependent changes in lipoxygenase and NADPH oxidase gene expression in *Arabidopsis thaliana* exposed to cadmium or excess copper. Funct Plant Biol 37:532-544

51. Santos RW, Schmidt ÉC, Martins RDP, Latini A, Maraschin M, Horta PA, Bouzon ZL (2012) Effects of cadmium on growth, photosynthetic pigments, photosynthetic performance,

biochemical parameters and structure of chloroplasts in the agarophyte *Gracilaria domingensis* (Rhodophyta, Gracilariales). Amer J Plant Sci 3:1077-1084

52. Sárvári É (2005) Effects of heavy metals on chlorophyll-protein complexes in higher plants: Causes and consequences. In: Pessarakli M (ed) Handbook of photosynthesis, USA pp: CRC Press, Boca Raton. pp 865-888
53. Sárvári É, Cseh E, Balczer T, Szigeti Z, Záray G, Fodor F (2008) Effect of Cd on the iron re-supply-induced formation of chlorophyll-protein complexes in cucumber. Acta Biol Szeged 52(1):183-186
54. Sheoran IS, Signal HR, Singh R (1990) Effect of cadmium and nickel on photosynthesis and the enzymes of photosynthetic carbon reduction cycle in pigeon pea (*Cajanus cajan* L.). Photosynth Res 23:345-351
55. Shukla UC, Murthy RC, Kakkar P (2008) Combined effect of ultraviolet-B radiation and cadmium contamination on nutrient uptake and photosynthetic pigments in *Brassica campestris* L. seedlings. Environ Toxicol 23:712-719
56. Siedlecka A, Baszyński T (1993) Inhibition of electron flow around photosystem I in chloroplasts of Cd-treated maize plants is due to Cd-induced iron deficiency. Physiol Plant 87:199-202
57. Sigfridsson KGV, Bernat G, Mamedoy F, Styring S (2004) Molecular interference of $Cd^{2+}$with photosystem II. Biochim Biophys Acta 1659:19-31
58. Skrebsky EC, Tabald LA, Pereira B, Rauber R, Maldaner J, Cargnelutti D, Gonçalves JF, Castro GY, Shetinger MRC, Nicoloso FT (2008) Effect of cadmium on growth, micronutrient concentration, and δ-aminolevulinic acid dehydratase and acid phosphatase activities in plants of *Pfaffia glomerata*. Braz J Plant Physiol 20(4):285-294
59. Somashekaraiah B, Padmaja K, Prasad A (1992) Phytotoxicity of cadmium ions on germinating seedlings of mung bean (*Phaseolus mungo*): involvement of lipid peroxides in chlorophyll degradation. Physiol Plant 85:85-89
60. Stobart AK, Griffith WT, Ameen-Bukhari J, Sherwood RP (1985)

The efffect of $Cd^{2+}$ on the biosynthesis of chlorophyll in leaves of barley. Physiol Plant 63:293-298

61. Tamas L, Dudikova J, Durcekova K, Haluskova L, Huttova J, Mistrik I (2009) Effect of cadmium and temperature on the lipoxygenase activity in barley root tip. Protoplasma 235:7-25
62. Timperio AM, D'Amici GM, Barta C, Loreto F, Zolla L (2007) Proteomics, pigment composition, and organization of thylakoid membranes in iron-deficient spinach leaves. J Exp Bot 58:3695-3710
63. UNEP (2010) Final review of scientific information on cadmium. DTIE: UNEP Chemical Branch.
64. Valentoviová K, Halušková L, Huttová J, Mistrík I, Tamás L (2010) Effect of cadmium on diaphorase activity and nitric oxide production in barley root tips. J Plant Physiol 167:10-14
65. Vassilev A, Lidon F, Scotti P, Da Graca M, Yordanov I (2004) Cadmium-induced changes in chloroplast lipids and photosystem activities in barley plants. Biol Plant 48:153-156
66. Wahid A, Arshad M, Farooq M (2009) Cadmium phytotoxicity: responses, mechanisms and mitigation strategies. In: Lichtfouse E (ed) Advances in sustainable agriculture, vol 1. Netherlands: Springer, pp 371-403
67. Wang H, Zhao SC, Liu RC, Zhou W, Jin JY (2009) Changes of photosynthetic activities of maize (*Zea mays* L.) seedlings in response to cadmium stress. Photosynthetica 47:277-283
68. Wang F, Chen F, Cai Y, Zhang G, Wu F (2011) Modulation of exogenous glutathione in ultrastructure and photosynthetic performance against Cd stress in the two barley genotypes differing in Cd tolerance. Biol Trace Elem Res 144(1–3):1275-1288
69. Wang X, Zhang ZW, Tu SH, Feng WQ, Xu F, Zhu F, Zhang DW, Du JB, Yuan S, Lin HH (2013) Comparative study of four rice cultivars with different levels of cadmium tolerance. Biologia 68:74-81
70. Wodala B, Eitel G, Gyula TN, Ördög A, Horváth F (2012) Monitoring moderate Cu and Cd toxicity by chlorophyll fluorescence and P700 absorbance in pea leaves. Photosynthetica 50:380-386

71. Xue XC, Gao HY, Zhang LT (2013) Effects of cadmium on growth, photosynthetic rate and chlorophyll content in leaves of soybean seedlings. Biol Plant 57(3):587-590

72. Ying RR, Qiu RL, Tang YT, Hu PJ, Qiu H, Chen HR, Shi TH, Morel JL (2010) Cadmium tolerance of carbon assimilation enzymes and chloroplast in Zn/Cd hyperaccumulator *Picris divaricata*. J Plant Physiol 167:81-87

73. Zhang M, Senoura T, Yang X, Chao Y, Nishizawa NK (2011) Lhcb2 gene expression analysis in two ecotypes of *Sedum alfredii* subjected to Zn/Cd treatments with functional analysis of SaLhcb2 isolated from a Zn/Cd hyperaccumulator. Biotechnol Lett 33:1865-1871

Chapter 6

# Integrative Medicine and Human Health - The Role of Pre-, Pro- and Synbiotics Integrative Medicine and Human Health - The Role of Pre-, Pro- and Synbiotics

Stig Bengmark

Division of Surgery & Interventional Science, University College London, 4th floor, 74 Huntley Street, London, WC1E 6AU, UK 185 Barrier Point Rd, London, E162SE, UK

## ABSTRACT

Western lifestyle is associated with a sustained low grade increase in inflammation -increased levels of endotoxin in the body and increased activation of Toll-like receptors and neutrophils, which leads to impaired immunity and reduced resistance to disease, changes which might explain the epidemic of chronic diseases spreading around the globe. The immune system cannot function properly without access to bacteria and raw plants, rich not only in bacteria but also in plant fibre, antioxidants, healthy fats and numerous other nutrients. Modern food technology with plant breeding, separation, condensation of food ingredients, heating, freezing, drying, irradiation, microwaving, are effective tool to counteract optimal immune function, and suspected to be a leading cause of so called Western diseases. Supply of pre-, pro-, and synbiotics have sometimes proved to be effective tools to counteract, especially acute diseases, but have often failed, especially in chronic diseases. Thousands of factors contribute to unhealth and numerous alterations in life style and food habits are often needed, in order to prevent and cure "treatment-resistant" chronic diseases. Such alterations include avoiding processed foods rich in pro-inflammatory molecules, but also a focus on consuming substantial amounts of foods with documented anti-inflammatory effects, often raw and fresh green vegetables and tubers such as turmeric/curcumin.

# REVIEW

## Introduction

Human life without access to plants and bacteria would be miserable. Plants and bacteria, which have existed for billions of years, have often robust protection system, which can be used by humans. Our Palaeolithic forefathers did on annual basis receive their daily food from at least five hundred plants and also, as the food they ate often were stored in the soil, a rich supply of various microorganisms. The food of modern food is based on nutrients received from only a small number of plants; 80% of the nutrients come from 17 plants and 50% of the calories from eight grains. Furthermore, the main part of Western foods is extensively processed; growth enhancement, separation, condensing, drying, freezing, irradiation, burning, microwaving, toasting, adding various ingredients and especially heating. It is well-known that some important plant ingredients start disappearing already when heated

pronounced effects of fibres on glycemic index are obtained by water-soluble fibres. Guar-gum is in this respect by far the most clinically tried fibre and will, as based on 15 different studies, reduce blood glucose to almost half (44%).

Lipid control/prevention of coronary heart disease. Soluble fibres such as pectin, guar gum, betaglucans (oat) reduce significantly blood cholesterol both in hypercholesterolemic and normocholesterolemic individuals, effects not found when non-soluble fibres such as cellulose and wheat bran are tried. Soluble fibres are excellent substrates for production in the large intestine of short chain fatty acids (SCFAs), known prevent leaky gut and also to reduce the levels of cholesterol/s.

A meta-analysis reports statistically significant protective effects against coronary heart disease in 14/16 studies [12]. In addition, larger fibre consumption is also known to reduce clotting and increase fibrinolysis, also important for prevention of arterial wall plaques and of thrombosis formation.

## Clinical Use of Fibres

Substances, important to health: amino acids such as arginine, glutamine, histidine, taurine, various sulphur and related amino acids, polyamines, omega-fatty acids, numerous vitamins and antioxidants are all to a great extent supplied to the body from plants. One cannot expect any significant amount of antioxidants to be delivered to the lower level of the gastrointestinal tract, if not carried by plant fibres. It is important to remember that key nutrients such as omega-3 fatty acids, glutamine, glutathione and several other nutrients are heat-sensitive and do not tolerate processing or storage to any larger extent. Plant fibres, which have been dried, heated up or micro-waved cannot be expected to contain any larger amounts of these, with this raw and unprocessed foods are needed. It is highly desirable that, whenever possible, the supply of commercial nutrition formulas are complemented by supply of fresh fruit and vegetable juices, as locally produced as ever possible. It is also desirable that several different fibres are supplied in parallel, and that both soluble and non-soluble fibres are supplied. For example, oat fibres are mainly metabolized in the proximal colon, while wheat fibres are known to be effective in the distal part of the colon, e.g. the part of colon where most cancers are localized. Oat has mainly shown sepsis-reducing effects while wheat has mainly been effective in cancer prevention.

Among the fibres commonly used in clinical nutrition are:

*Algal fibres.* Most of the algal fibres are resistant to hydrolysis by human endogenous digestive enzymes, but are to various degrees fermented by colonic flora. The soluble fibres consists in lamarans (a sort of -glucan associated with mannitol residues), fucans (sulphated polymers associated with xylose, galactose and glucoronic acid) and alginates (mannuronic and guluronic acid polymers). The insoluble algal polymers consist mainly in cellulose. Fermentation of alginates yields high levels of acetate (80%), while lamirans yield preferably butyrate (16%). It is most likely that algal fibres will within a few years be routinely used in clinical nutrition.

*Fructans.* Fructans starches and sucrose serve in the plant as its energy reserve. These substances are also produced by bacteria and fungi. Fructans are said to enhance the tolerance of the plant to stressful conditions and make it possible for the plants to survive under harsh conditions, such as low temperature and draft. The most well-known fructans are inulin (rich in chicory, artichoke, onions, and banana) and phleins (rich in various grasses). This far it is mainly inulin, which has been tried in human nutrition. Various oligosaccharides are reported to stimulate the flora and especially the growth of *Lactobacilli* and *Bifidobacteria*in the large intestine and to reduce the content of potentially pathogenic microorganisms (PPMs) in the intestine. Increase in the *Bifidobacteria* flora is regarded as especially favourable as *Bifidobacteria* are known to produce important vitamins, among them thiamine, folic acid, nicotinic acid, pyridoxine and vitamin $B_{12}$, which is of great importance for health. A fructan called neokestose, found in onion, is reported to have even better ability than inulin to promote growth of LAB. Also supplementation of fructans is reported to reduce concentrations in serum of insulin, cholesterol and triacylglycerol. It is also reported to promote absorption of calcium and other minerals. Other oligosaccharides such as those extracted from peas and beans especially soya bean oligosaccharide (raffinose and stachyose) and pyrodextrin, produced by pyrolysis of maize and potato starch are also reported to be beneficial for human health.

*Glycomannans.* Glycomannan, a glucose/mannose polymer derived from a plant called*Amorphophallus konjak*, which has several

English names such as devil tongue, elephant yam and umbrella arum. It has unique hydroscopic abilities and will on contact with water swell and form a viscous gel, which like other gels will delay gastric emptying and intestinal transit time. It has been shown to be effective in delaying absorption of digestible energy. It has this far mainly been used in Japan and other Asian countries to treat diabetes, hypertension and hypercholesterolemia. Dietary supply of konjak mannans has been shown in experimental animals to alter the flora and reduce tumorigenesis. It is also effective to control of diarrhoea in enteral nutrition, especially in elderly patients, and to increase the *Bifidobacteria* flora.

*Oat gum.* Oat contains a series of interesting compounds, which is the reason why an increasing part the world production of oat goes to the pharmaceutical and cosmetic industries. The amino acid pattern of oat is rather similar to that of human muscle (only that of buckwheat is more alike), and can thus be expected to deliver most of the amino acids needed to build muscles. Oat is rich in water-soluble fibers, β-glucans, and known for their antiseptic properties. Oat is also rich in natural antioxidants, particularly ferulic acid, caffeic acid, hydrocinnamic acid, and tocopherols and oat was before synthetic antioxidants were available extensively used to preserve foods: milk, milk powder, butter, ice-cream, fish, bacon, sausages and other food products sensitive to fat oxidation. Another ingredient richly available in oat is inositol hexaphosphate (phytic acid), a strong antioxidant, particularly known to enhance natural killer cell activity and to suppress tumour growth. Oat is also rich in polyunsaturated fats/polar lipids such as phosphatidylcholine, known for its protective effects of mucosal and cellular surfaces.

*Pectin.* Also pectin is an interesting fibre, extensively used by pharmaceutical and food industry. It has a unique ability to form gels and is commonly used as a carrier of pharmacologically active substances and is common in baby foods. An important finding is that pectin a very strong antioxidant against the three most dominating oxidation damages induced by peroxyl-, superoxide- and hydroxyl radicals. These effects might explain why pectin has the capacity to stimulate the gut-associated immune system and to prevent disruption of the intestinal microflora. Pectins have shown strong protective and healing effects on gastric but also intestinal mucosa, in experimental studies not inferior to what is observed with $H_2$-blockers, proton inhibitors and surface-protection agents [13, 14]. Pectin builds a protection layer in the stomach and facilitates maintenance of gastric acidity, important for prevention of colonization of the stomach by pathogens. Pectin is also an excellent substrate for microbial fermentation.

## Lactic Acid Bacteria Key to Fermentation of Fibres

Not all fibres are easily fermented in the gut. Among the more fermentation-resistant fibres are wheat fibres which usually are not digested until they reach descending colon. Also oligofructans (inulin or phleins) are difficult to ferment and only a small minority of lactic acid bacteria (LAB) are able to do so. When the ability of 712 different LAB to ferment oligofructans was studied, only 16/712 were able to ferment the phleins and 8/712 inulin [15]. Apart from*Lactobacillus plantarum* only three other LAB species, *Lactobacillus paracasei* subsp. *paracaseiLactobacillus brevis* and *Pediococcus pentosaceus* were able to ferment such semi-resistant fibres. Another study investigated the ability of 28 different LAB to ferment pure fructooligosacharides (FOS). All *Lb plantarumLb casei* and *Lb acidophilus* strains studied and most *Bifidobacteria* metabolized FOS, in contrast to yoghurt bacteria such as *Lb bulgaricus* and*Streptococcus thermophilus* and also *Lactobacillus strain GG,* which all were unable to ferment these fibres [16].

## Clinical Experience with Supplemented Plant Fibres

### *Plant fibres in constipation*

Chronic constipation is one of the most common disorders in Western countries. Its etiology remains despite numerous clinical, pathophysiologic, and epidemiologic studies unclear, but it is suggested high intake of dairy products and intake of plant fibres play a significant role in its pathogenesis. A randomized sample of 291 children with idiopathic chronic constipation was in a case control study compared with 1602 healthy controls [17]. Constipation was clearly negatively correlated with low intake of cellulose and pentose fibres ($p < 0.001$).

Fructo-oligosaccharides (FOS) may also have potential benefits in constipation, since they exhibit many soluble dietary fibre-like properties. In a study a total of fifty-six healthy infants, age 16–46 weeks (mean age 32 weeks) were randomly assigned to receive either 0.75 g FOS or placebo added to a serving of cereals for 28 d [18]. The mean number of stools per infant was 1.99 ± 0.62 per day in

## Plant fibre to prevent and treat diarrhoea

In a large randomized study in acutely ill medical and surgical patients, all requiring enteral nutrition for a minimum of 5 days supplementation of hydrolyzed guar gum was compared to a fibre-free enteral nutrition, the incidence of diarrhoea being 9% with fibre-supplementation, compared to 32% with fibre-free nutrition ($p > 0.05$) [19]. One of the effects of certain fibres is that it increases the bioavailability and absorption of zinc, which is especially shown for oligosaccharides. Zink supplementation was in a randomized study in 3–59 months children in Bangladesh proven effective to lower both the incidence of diarrhoea and the duration of diarrhea [20]. In another study from Bangladesh 250 g/L of green (unripe) banana (eqv. to two fruits) or 2 g pectin/kg food was compared to a supplemented rice diet in children suffering from persistent diarrhoea [21]. The amounts of and frequency of stools, the duration of diarrhoea, numbers of vomiting, use of oral rehydration and amounts iv fluid solutions given were all significantly better reduced in the the group supplemented green banana and pure pectin. Recovery on third day was seen in 59% in the green banana group, in 55% in the pectin group compared to 15% in the rice only control group.

## Plant fibre to support mineral absorption

It is well accepted that nutrition is of great importance also for bone health. Most interest has this far focused on calcium and vitamin D. Much less interest has been paid to other important nutrients such as protein, and especially to minerals such as phosphorus, potassium, magnesium and vitamins such as C and K. Recent studies suggests that increased intake of plant fibers, fruits and vegetables is associated with an increased bone mineral density also in elderly subjects, both women and men [22, 23]. Of all the many pure fibres available it is mainly the effects of oligosaccharides have been more thoroughly studied, but so far mainly in experimental animals Calcium absorption, bone calcium content, bone mineral density, bone balance and bone formation/bone absorption index are generally reported to be significantly increased, and that as early already as after three weeks of supplementation of a mixture of inulin and fructooligosaccharides.

## Plant fibre to control weight

No major effects on body weight by supplementation of only prebiotic fibre have this far been reported. The effects of dietary fibre on subjective hunger ratings and weight losses were studied some twenty years ago in members of a weight loss club. One hundred and eight of 135 members completed the trial: 23 controls, 45 on ispaghula granulate and 40 on bran sachets [24]. Both fibre preparations reduced hunger at all meals. The mean (±SD) weight reductions during the trial were $4.6 \pm 2.7$ kg for the controls, $4.2 \pm 3.2$ kg for the ispaghula group and $4.6 \pm 2.3$ kg for the bran group ($p > 0.05$ for both groups). Although supply of dietary fibre immediately before meals did reduce the feeling of hunger did it not provide any additional benefits to the weight reduction. A more recent crossover study compared the effect on satiety of supplementation of $27 \pm 0.6$ g/d of fermentable fibers (pectin, betaglucan) with similar amounts of non-fermentable fibre (methylcellulose). The daily satiety was significantly increased with non-fermentable (methylcellulose) than with fermentable fibres (betaglucan, pectin) ($p = 0.01$) but no differences were observed in daily energy intake or loss of body weight or body fat [25].

## Plant fibre in inflammatory bowel diseases (IBD)

Although both patients with IBD and irritable bowel syndrome (IBS) are known to under-consume dietary fibres, there is little evidence that lack of dietary fibre plays a role in the pathogenesis of these diseases. The ability of maintaining remission in UC patients by a daily supply of 10 g of Plantago ovata seeds (also called psyllium or Ispaghula husk) was compared with daily treatment with 500 mg of mesalamine and a combination of the two [26] Twelve months of treatment failed to demonstrate any difference in clinical benefits between the three groups. Germinated barley foodstuff (GBF), a by-product from breweries, rich in hemi-cellulose and in glutamine, was tried in 39 patients with mild-to-moderate active UC [27]. Daily supply of 30 g reduced significantly the disease activity, increased concentration of SCFAs, and increased in stool the numbers of *Bifidobacterium and Eubacterium*. It can well be that the observed effect were more due to increased supply of glutamine and other antioxidants such as various B vitamins than to the fibre per se as these compounds are known to be rich in by-products from breweries. Glutamine, as well as other antioxidants, are known to attenuate pro-inflammatory cytokines such as TNF- and to enhance release of heat shock proteins (HSP-72) [28].

A controlled study using oat bran as fibre source was reported in a study of 22 patients and 10 controls, all with quiescent UC. Daily supply during three months of as much as 60 g of oat bran (eqv to 20 g dietary fibre) resulted in a significant increase in faecal butyrate (average 36%), but also in significant reduction in abdominal pain. All the treated patients tolerated well the large dose of fibre and signs of relapse of disease were seen in none of the colitis patients [29]. Butyrate has been shown to inhibit NF-kB activation of lamina propria macrophages, and to reduce the number of neutrophils in crypts and surface epithelia, as well as the density of lamina propria lymphocytes/ plasma cells in patients with ulcerative colitis [30] - findings correlating well with the observed decreased disease activity. Twenty patients with ileal pouch-anal anastomosis received daily for two weeks 24 g of inulin. Significant reduction in inflammation was observed at endoscopy and histology. In addition significant increase in faecal concentrations of butyrate and reductions in faecal pH, faecal content of secondary bile acids, and growth of *Bacteroides fragilis* was observed [31].

## Plant fibre in irritable bowel disease

Dys-motility disorders are increasingly common in Western Societies. Some evidence suggest that various dys-motility disorders; gastroesophageal reflux problems, infant colic and constipation are all food-related features, and often due to intolerance to cow's milk proteins[32]. Irritable bowel syndrome (IBS), is a clinical diagnosis based on the occurrence of abdominal distension, abdominal cramps, often increased transit time, more frequent stools, and relief of pain on defecation. The prevalence of the syndrome varies between and 7 and

22%, making IBS the most common functional gastro-intestinal disorder [33]. Unfortunately, no effective pharmaceutical treatment exists, or when existing, is unacceptably toxic [34]. This has resulted in a need for additional modalities for the treatment of IBS. Pre- and probiotics appear in this perspective as attractive alternatives, see further reviews [35,36] especially as early data from human intervention studies but also results from experimental studies in animals clearly demonstrate an impact on the immune system: immune cells of the GALT including Peyer's patches [37]. However, a consequence of feeding the currently favoured prebiotics (inulin, fructo-oligosaccharides (FOS), trans-GOSs and lactulose,) is increased gas production in the gut, which might preclude prebiotic use where bloating or gas are prominent symptoms such as in diarrhoea-predominant IBS, but might allow their mild laxative properties to be useful in constipation-predominant IBS [38]. Over the years some small open trials have been performed, but this far, no larger and randomized trial has been reported. However, a small open label trial with supplementation of 15 g/day of a mixture of oligofructose (70%) and inulin (30%) reported significant reduction in disease activity (Harvey Bradshaw index fell from 9.8, SD 3.1 to 6.9 SD 3.4, $p = 0.01$) in parallel to a significant increase in faecal bifidobacteria concentration (from 8.8, SD 0.9 log 10 to 9.4, SD 0.9 log10 cells/g dry feces $p = 0.001$). Also the IL-10 positive dendritic cells increased (from 30 to 53% $p = 0.06$), and the percentage of dendritic cells expressing TLR2 and TLR4 increased from 1.7% to 36.8% $p = 0.08$, and from 3.6% to 75.4 $p = 0.001$) [38], respectively which offers hope for the future.

## Plant fibre in abdominal pain

Other dietary fibres have also been tried in various groups of abdominal pain. A recent Cochrane review was unable to find any evidence that fibre supplements, lactose free diets or lactobacillus supplementation are effective in the management of children with recurrent abdominal pain [39]. However, a study in adult patients reports significant success with other fibres than classical prebiotics. 188 adult IBS patients were classified as having diarrhoea-predominant, constipation-predominant, or changeable bowel habits type IBS and randomly assigned to groups receiving 30 g/day of wheat bran or 5 g/day of guar gum (PHGG [40]. After four weeks, patients were allowed to switch group, depending on their subjective evaluation of their symptoms. Both fibre and PHGG were effective in improving pain and bowel habits. Significantly more patients switched from fibre to PHGG (49.9%) than from PHGG to fibre (10.9%) at four weeks. Intention-to-treat analysis showed a significantly greater success in the PHGG group (60%) than in the fibre group (40%). In addition, significantly more patients in the PHGG group reported a greater subjective improvement than those in the fibre group. It was concluded that improvements in core IBS symptoms were observed with both bran and PHGG, but the latter was better tolerated and preferred by patients [40].

*Plant fibre to control infections.* In an effort to prevent nosocomial pneumonia and sepsis, patients with severe multiple trauma were treated with beta 1–3 polyglucose (glucan) - a component of cell walls of plants and microbes [41]. Pneumonia occurred in 2/21 glucan-treated and in 11/20 patients in the control group ($p<0.01$). Infectious complications (pneumonia and/or general sepsis) occured in 14% of glucan-supplemented patients versus 65% in the control group ($p<0.001$). Another study compared the effects of a high-protein formula enriched with fiber but also arginine, and antioxidants with a standard high-protein formula in early enteral nutrition in critically ill patients [42]. The supplemented group had, in comparison to non-supplemented controls a lower incidence of catheter-related sepsis (0.4 episodes/1000 ICU days) than the control group (5.5 episodes/1000 ICU days) ( $p<.001$), but no differences were observed between the groups in incidence of ventilator-associated pneumonia, surgical infection, bacteremia, urinary tract infections, mortality and in long-term survival [42].

## Pre- pro- and synbiotics improve innate immunity

Several plant fibres (prebiotics) and a few LABs (probiotics) have documented significant effects to improve both the function of the innate immune system, the physical barrier, and increase resistance to disease. The hope is that combined supply of these components shall have synergistic, e.g. more than additive effects in boosting the immune system and enforcing the barrier functions. Products which combine pre-and probiotics are called synbiotics and treatments using the combination called synbiotic treatment.

## Choice strains for probiotic use

The choice of pre- and probiotics must be based on scientific evidence. Stronger bioactivities cannot be expected from LAB such as yogurt bacteria, chosen mainly for their palatability. The strains to use must be selected with care and based on extensive preclinical.

It is important to remember that the majority of LABs have no or much limited effects on immune functions and outcome. Constructing synbiotic formulations is especially demanding as most of the LAB used by industry have no or limited ability to ferment bioactive fibres such as inulin or phlein [15], no ability to adhere to human mucus, low antioxidant capacity and most important, do not survive the acidity of stomach and bile acid content.

It has been observed that strains that carry the same name have different and sometimes opposing functions. When the ability of 46 different *Lactococcus lactis* strains to induce production of the cytokines interleukin (IL)-6, IL- 12 and tumor necrosis factor (TNF)- was studied, it was clearly demonstrated that the extent of induction of IL-6, IL-12 and TNF- was cearly shown to be strain-specific and not at all related to subspecies, bio-variety, or the source of the isolate. The production of IL-6 varied between 138 and 0 ng/ml IL-12 between 3 and 0 ng/ml TNF- between 20 and 0 ng/ml between various strains carrying the same name [43].

Unfortunately few studies have looked at the synergistic effects of simultaneous supply of LAB and fibres - synbiotics. Natural foods supplies both LAB and a great variety of plant fibres. Combination of several fibres has been shown to lead to additive effects on microbial ecosystem and immune responses [44], and multi-species probiotics documented to be superior to single-species probiotics, showing increased ability to enhance growth, reduce antibiotic associated diarrhoea, prevent infections (*S typhimurium*) and reduce pathogenic colonisation (*E coli*) [45] Although some studies have used various synbiotic compositions, only two such compositions have been produced after extensive preclinical studies:

## An Unistrain/Unifibre Composition

, produced (Probi AB, Lund Sweden) by fermentation of oat meal with *L. plantarum* strain 299, containing $10^9$ of LAB and app 10 g oat fibre [46]. In a few studies a commercial fruit juice, PRO VIVA™ containing $10^7$ of a related *L. plantarum* strain called 299 V, (Skånemejerier, Malmö, Sweden) is also tried.

## A Multistrain/Multifibre Composition

, called Synbiotic 2000™, consisting in a mixture of $10^{10}$ and a Synbiotic FORTE™ with $10^{11}$ of each of four LAB: *Pediacoccus pentosaceus* 5–33:3, *Leuconostoc mesenteroides* 32–77:1,*Lactobacillus paracasei* subsp paracasei 19, and *Lactobacillus plantarum* 2362 and 2.5 g of each of the four fermentable fibres (prebiotics): betaglucan, inulin, pectin and resistant starch (Synbiotic AB, Höganäs, Sweden) [47,48].

Lund University microbiologists Åsa Ljungh and Torkel Wadström developed this multi-strain/multi-fibre synbiotic formula, which in recent years has been extensively used in clinical trials. The choice of LAB for the formulation was done after extensive studies of > 350 human[47] and >180 plant microbial strains [48] and based especially the ability of the LAB to produce bioactive proteins, transcribe NF- B, produce pro- and anti-inflammatory cytokines, produce antioxidants, and most important, to functionally complement each other. In recent studies both the Synbiotic 2000 FORTE™ but also a Probiotic 2000 FORTE™ (no fibre added), containing $10^{11}$ of each of the four LAB, e.g. 400 billion LAB per dose have been tried.

## Plantarum, paracasei and pediococcus

When, as mentioned above, the ability of 712 different LAB to ferment oligofructans was studied, only 16/712 were able to ferment semi-resistant fibred; the phleins and 8/712 inulin;*Lactobacillus plantarum, Lactobacillus paracasei* subsp. paracasei*Pediococcus pentosaceus* and*Lactobacillus brevis* and were able to ferment these semi-resistant fibres [15]. Interesting clinical results are also often obtained when these LAB are involved. When more than 100 LAB were compared *Lb paracasei* subsp *paracasei* were demonstrated to be the strongest inducer of Th1 & repressor of Th2 cytokines [49]. Several other studies have also documented the unique ability of *Lb paracasei* to iinduce cellular immunity, stimulate production of suppressive cytokines as TGF and Il-10, suppress CD4 T-cells, suppress Th2 activity, suppress splenocyte proliferation and decrease antigen-specific IgE and IgG1 [50-53].

The effect of *Lactobacillus paracasei* (NCC 2461), *Lactobacillus johnsonii* (NCC 533) and*Bifidobacterium lactis* Bb12 (NCC 362) on the inductionandmaintenanceoforaltolerancetobovinebeta-lactoglobulin (BLG) was investigated in mono-colonized germfree mice. The effects of *L. paracasei* were reported superior to those of the other two [53]. A study, which compared the ability of 50 different LAB to control 23 different pathogenic *Clostridium difficile* found more than half (27/50) totally ineffective, (18/50) antagonistic to some , but only five strains effective against all: two strains of *Lb paracasei* s. *paracasei* and five strains of *Lp plantarum*[54]. Another study compared the effects in rats receiving during days 10 to 21 after*Trichinella spiralis* - induced infection either *Lactobacillus paracaseiLactobacillus johnsoniiBifidobacterium longum*, or *Bifidobacterium lactis; Lb paracasei* but not the other LAB attenuated muscle hyper-contractility, reduced the infection-associated Th- 2 response and muscle levels of TGF- , COX-2 and PGE2 [55]. A recent study compared in animals the effects of three probiotic strains: *Bifidobacterium lactis* NCC362, *Lactobacillus johnsonii* NCC533, and*Lactobacillus paracasei* NCC2461 on stress-induced changes in gut permeability & on sensitivity to colorectal distension (CRD) Only *Lb paracasei* both prevented reduced significantly existing visceral pain and also restored normal gut permeability [56].

## ***Synbiotic 2000 in clinical medicine***

I have dedicated most of my clinical research during the last two decades to studies on effects of synbiotic compositions in various clinical situations, during the 1990s to the monostrain/monofibre composition, mentioned above, and 2000s to the multistrain/multifibre composition, also mentioned above; Synbiotic 2000 and 2000 Forte res. Here follows a summary of the effects observed this far. It has been tried in the following conditions:

## Acute pancreatitis

Sixty-two patients with severe acute pancreatitis (SAP) (Apache II scores: Synbiotic 2000-treated 11.7 ± 1.9, controls 10.4 ± 1.5) were given either two sachets/day of Synbiotic 2000™(2x40 billion LAB/day and totally 20 g fibres) or the same amounts of fibres in (20 g) as in Synbiotic 2000™ during the first 14 days after arrival to the hospital [57]. 9/33 patients (27%) in the Synbiotic 2000-treated group and 15/29 patients (52%) in the only fibre-treated group developed subsequent infections. 8/33 (24%) Synbiotic 2000-treated and 14/29 (48%) of the only fibre-treated patients developed SIRS, MOF or both ($p < 0.005$). A total of pathogenic microorganisms of seven were cultivated in the Synbiotic treated group compared to seventeen in the fibre-only group (Figure 1).

| Isolated bacteria: | Synbiotic | Fibres only |
|---|---|---|
| ◆ Enterococcus faecalis | 1 | 11 |
| ◆ Escchericha coli | 0 | 3 |
| ◆ Enterobacter cloacae | 0 | 2 |
| ◆ Pseudomonas aeruginosa | 0 | 2 |
| ◆ Staphylococcus aureus | 0 | 1 |
| Total | 1 | 18 |

**Figure 1:** Isolated bacteria in acute pancreatitis treated with Synbiotic 2000 or only the fibres.

## Polytrauma

Two prospective randomized trials one with Synbiotic 2000 and one with Synbiotic 2000 Forte respectively have been concluded. The first study compared in patients with acute extensive trauma: 1. Synbiotic 2000 (40 billion LAB/d) with 2. A soluble fibre 3. A peptide diet (Nutricomp, Braun Inc Germany) and 4. Supplementation of glutamine. Treatment with Synbiotic 2000™ lead to a highly significant decrease in number of chest infections (4/26 patients - 15%), compared to peptide diet (11/26 patients - 42%, $p<0.04$), glutamine (11/32 patients - 34%, $p<0.03$) and only fibres (12/29 patients- 41%, $p<0.002$) [58]. Also the total number of infections were significantly decreased; Synbiotic 2000™ 5/26 patients (19%) only fibres 17/29 patients (59%) peptide 13/26 patients (50%) and glutamine16/32 patients (50%).

In the second study sixty-five patients polytrauma patients were randomized to receive once daily for 15 d Synbiotic 2000 Forte (400 billion LAB + 10 gram of fibres, see above) or maltodextrine as placebo. Significant reductions were observed in number of deaths (5/35 vs 9/30, $p<0$, 02), severe sepsis (5/35 vs 13/30, $p<0.02$), chest infections (19/35 vs 24/30, $p<0.03$), central line infections (13/32 vs 20/30, $p<0.02$), and ventilation days (average 15 vs 26 days) [59]. A total of pathogenic microorganisms of 54 were cultivated in the Synbiotic treated group compared to 103 in the placebo group. The time to progression to primary bacteraemia was longer among patients treated with Synbiotic 2000 Forte compared with placebo (*p* _ 0.0237 between groups). Twelve (33.3%) and five (13.9%) placebo-treated and synbiotic-treated patients, respectively, developed ventilator-associated pneumonia with *Acinetobacter baumannii* as a bacterial cause (*p* 0.047 between groups). Treatment with Synbiotic 2000 Forte was accompanied by reduction of white blood cell counts and LPS and CRP levels in either patients who did or did not develop sepsis [60].

## Abdominal surgery

In a randomized controlled study forty-five patients undergoing major surgery for abdominal cancer were divided into three treatment groups: 1. enteral nutrition (EN) + Synbiotic 2000 (LEN), 2. EN + only the fibres in the same amounts (20 g) (20 g) as in Synbiotic 2000™ (FEN) and 3. A standard parenteral nutrition (PN). All treatments lasted for 2 preoperative and 7 days postoperative days. The incidence of postoperative bacterial infections was 47% with PN, 20% with FEN and 6.7% with LEN ($p<0.05$) (Han Chun Mao et al in press). A total of pathogenic microorganisms of 34 were cultivated in the Synbiotic treated group compared to 54 in the fibre-only group. Significant improvements were also documented in prealbumin (LEN, FEN), C-reactive protein (LEN, FEN), serum cholesterol (LEN,FEN), white cell blood count (LEN) , serum endotoxin (LEN, FEN) and IgA (LEN).

In another prospective randomized double-blind trial performed in 80 patients subjected to pylorus-preserving pancreatoduodenectomy (PPPD) receive twice daily either Synbiotic 2000™ (2x40 billion LAB) or only the fibres in composition from the day before surgery and during the first seven postoperative days [61]. A highly significant difference in infection rate ($p = 0{,}005$) was observed as only 5/40 patients (12.5%) in the Synbiotic 2000-treated group suffered infections (4 wound and one urinary tract infection) versus 16/40 (40%) in the Only fibre group (6 wound infections, 5 peritonitis, 4 chest infections, 2 sepsis, and one of each of urinary tract infection, cholangitis and empyema). The infecting microorganism were in the Synbiotic treated group *Klebsiella pneumoniae* (2 patients),*Enterobacter cloacae* (2 patients), *Proteus mirabilis* (1 patient) and *Enterococcus faecalis/faecium* (1 patient) and in the Only fibre group *Enterobacter cloacae* (8 patients),*Enterococcus faecalis/faecium* (7 patient), *Escherichia coli* (7 patient), *Klebsiella pneumoniae* (2 patients), *Staphylococcus aureus* (2 patients), and *Proteus mirabilis* (1 patient) - see further Figure 2. Statistically significant differences between the groups were also observed in use of antibiotics (mean: Synbiotic 2000; 2 ± 5 days, Only fibres; 10 ± 14 days).

| Isoland bacteria | Synbiotic 2000 | Fibres only |
|---|---|---|
| • Enterobacter cloacae | 2 | 8 |
| • Enterococcus faecalis/faecium | 1 | 7 |
| • Escherichia coli | 0 | 7 |
| • Klebsiella pneumoniae | 2 | 2 |
| • Protenus mirabilis | 1 | 1 |
| • Staphylococcus ayreus | 0 | 2 |
| Total | 6 | 27 |

**Figure 2:** Isolated bacteria in patients undergoing pancreatoduodenectomy and treated with Synbiotic 2000 or Only the fibres.

## *Chronic liver disease and liver transplantation*

Fifty-eight patients with liver cirrhosis suffering so called minimal encephalopathy were randomized into three treatment groups: Group

1 (20 patients) received Synbiotic 2000 (40 billion LAB), group 2. (20 patients) received the same amount of the fibres in Synbiotic 2000 and group 3. (15 patients) received placebo (non-fermentable, non-absorbable fibre -crystalline cellulose) [62]. A significant increase in intestinal LAB flora was observed after one month of supplementation in the Synbiotic-treated group, but not in the other two groups. Intestinal pH was significantly reduced in both treatment groups but not in the placebo-treated group. Significant decreases in faecal counts of *Escherichia coli, Staphylococcus* and *Fusobacterium,*but not in *Pseudomonas* and *Enterococcus,* and significant decreases in ammonia/s, endotoxin/s ALT/s and bilirubin/s (original level 252 ± 182) were observed in the Synbiotic 2000-treated group ( 84 ± 65, $p < 0.01$) and in the Only fibre-treated group (110 ± 86, $p < 0.05$) while it remained unchanged in the placebo group. The improvements in liver function were accompanied by significant improvements in psychometric tests and in degree of encephalopathy.

In a follow-up study by the same group of investigators 30 patients with liver cirrhosis were randomized to receive either Synbiotic 2000 or placebo (crystalline cellulose ) for 7 days [63]. Viable fecal counts of *Lactobacillus species,* Child-Pugh class, plasma retention rate of indocyanine green ($ICG_{R15}$), whole blood tumour necrosis factor alpha (TNF--6 (IL-6) mRNA, serum TNF- , soluble TNFreceptor (sTNFR)I, sTNFRII and IL-6 and plasma endotoxin levels were measured pre- and post-treatment: Synbiotic treatment was associated with significantly increased faecal lactobacilli counts and significant improvements in plasma retention rate of indocyanine green ($ICG_{R15)}$ and stage of liver disease (Child-Pugh classification). No significant changes in any study parameter followed placebo treatment, but significant increases in whole blood whole blood tumour necrosis factor alpha (TNF- ) mRNA and interleukin-6 (IL-6) mRNA, along with serum levels of soluble TNF receptors sTNFRI and sTNFRII, were observed in the Synbiotic 2000-treated patients. TNF- and IL-6 levels correlated significantly, both at baseline and post-synbiotic treatment. Synbiotic-related improvement in $ICG_{R15}$ was significantly associated with changes in IL-6, both at mRNA and protein levels, and unrelated to plasma endotoxin values. It was concluded that even short-term synbiotic treatment can significantly modulates gut flora and improve liver function in patients with cirrhosis. The observed benefits seemed unrelated to reduction in endo-toxaemia, but could be mediated, at least in part, by treatment-related induction of IL-6 synthesis by TNF- . These results offers great hope that synbiotic treatment to patients on waiting list for liver transplantation might prevent septic episodes, improve liver function, and promote successful outcome of surgery.

Sixty-six patients were randomized to either receive Synbiotic 2000 or only the fibres in Synbiotic 2000 in connection with human ortotopic liver transplantation. The treatment started already on the day before surgery and continued for 14 days after surgery. During the first postoperative month did only one patient in the Synbiotic 2000-treated group (3%) show signs of infection (urinary infection) compared to 17/33 (51%) in the patients supplemented with only the four fibers [64]. The infecting organisms were in the Synbiotic-treated group*Enterococcus fecalis* in 1 patient and in the only fibre-treated group *Enterococcus fecalis/fecium*11, *Escherichia coli* 3, *Enterobacter cloacae* 2, *Pseudomonas aeruginosa* 2 and *Staphylococcus aureus* in 1 patient, see further Figure 3. The use of antibiotics was in average 0.1 ± 0.1 d in the synbiotic-treated patients and 3.8 ± 0.9 d in the only fibre-treated group.

| Isolated bacteria: | Synbiotic | Fibres only |
|---|---|---|
| ◆ Enterococcus faecalis | 1 | 11 |
| ◆ Escchericia coli | 0 | 3 |
| ◆ Enterobacter cloacae | 0 | 2 |
| ◆ Pseudomonas aeruginosa | 0 | 2 |
| ◆ Staphylococcus aureus | 0 | 1 |
| 'Total | 1 | 18 |

**Figure 3:** Isolated bacteria in patients undergoing liver transplantation and treated with Synbiotic 2000 or Only the fibres.

## *Inflammatory bowel disease*

Daily rectal instillations with Synbiotic 2000 reconstituted in saline were given to ten patients with distal colitis during 2 weeks. One patient withdraw after one week, the remaining patients showed during the 3 weeks of observation dramatic improvements in various disease scores; episodes of diarrhoea ((2.4 ⇒ 0.8), visible blood in stool (2.2 ⇒ 0.8), nightly diarrhoea (0.5 ⇒ 0), urgency (1.9 ⇒ 1.0) and consistency of stool ((1.1 ⇒ 0.8) [65]. Two patients reported significant bloating and wind but no other adverse or side effects were reported.

## Treatment-Resistant Conditions

Treatment with synbiotics have not this far improved the disease in two groups of conditions:

## Inflammatory bowel disease - Crohn's Disease (CD):

Sixty-three patients were after an initial treatment with infliximab randomized to daily receive either Synbiotic 2000 or crystalline cellulose as placebo [66]. Median time to relapse was 9.8 and 10.1 months respectively. In a second study were following surgery patients supplemented with either Synbiotic 2000 or crystalline cellulose as placebo. Seven patients in the synbiotic-treated group and two in the placebo group completed the scheduled 24 month treatment [67]. No differences were observed between the two groups either in endoscopic findings or rate of clinical relapse. The so called Rutgeerts disease scores were after three months of treatment $0.6 \pm 0.8$ in the synbiotic-treated group and $0.8 \pm 1$ in the placebo group (NS).

## General intensive care patients:

Two large studies have been performed in a general intensive care population; one with Synbiotic 2000 and one with Synbiotic 2000 FORTE. Synbiotic 2000 (40 billion LAB) was given to 162 patients and only fibre the fibres in the synbiotic composition to 168 patients. No difference was observed in mortality or in multi-organ dysfunction [68]. In another study Synbiotic 2000 FORTE 130 patients were supplemented twice a day throughout the whole ICU stay to ($2 \times 400$ billion LAB) and compared to 129 patients supplemented a cellulose based placebo. No statistical difference was demonstrated between the groups in the incidence of VAP (9 and 13%, $P = 0.31$). The rate of ventilator-associated pneumonia (VAP) per 1000 ventilator days was 13 and 14.6 ($p = 0.73$) and hospital mortality 27 and 33%, ($p = 0.32$), respectively[69].

# CONCLUSIONS

Thousands of factors are important to maintain health and to cure disease. This might explain why single drug pharmacy fails, both to prevent disease and to cure disease especially when chronic. Human innate immunity is for a proper function much depending on continuous access to bacteria and plants. Using probiotics in combination with plants and their active ingredients remains an attractive approach for prevention and treatment of various both acute and chronic diseases. Ten year old studies in the United States demonstrate an 83 percent reduction in rate of coronary heart disease [70], a 91 percent reduction in diabetes in women [71], and a 71 percent reduction in colon cancer in men [72] in patients adhering to what is regarded as an "healthy lifestyle": no use of tobacco, moderate use of alcohol, regular physical exercise, and eating a diet low in animal fat, low in refined carbohydrates, and rich in fresh fruits and vegetables (if raw also rich in lactic acid bacteria) and fish, a diet often referred to as Mediterranean diet.

Crohn's disease, as an example, has despite vigorous attempts over the years remained most resistant to therapy, pharmaceutical as well as probiotic treatments -see recent review [73]. Solid observations suggest that several Th1-mediated autoimmune diseases, such as multiple sclerosis, type 1 diabetes, rheumatoid arthritis and Crohn´s disease are associated with low vitamin D status, and a recent study demonstrates strong molecular links between vitamin D deficiency and the genetics of Crohn's disease [74]. Crohn's disease is also strongly associated with lack of minerals, especially calcium and magnesium. Hypomagnesaemia is known to be strongly associated with increased systemic inflammation, manifesting in leukocyte and macrophage activation and increased production of inflammatory cytokines and acute phase proteins. A recent study demonstrated a much deranged microbionta in experimental animals with induced hypomagnesemia [75]. It is certainly too much to request that pre-, pro-, or synbiotics shall make a difference in such conditions unless the underlying defects in metabolism and immune functions are corrected.

Inflammation is "the mother of disease" and much associated with the food we eat [76,77]. As pointed out above, numerous factors contribute to the deranged innate immune system and the sustained exaggerated inflammation. Numerous changes are also required to permanently control the hyper-inflammation and subsequent disease. Supply of pre-, pre-, and synbiotics are strong tools for such corrections of immune functions and resistance to disease. However, in most instances also other measures are necessary, including reduction in intake of pro-inflammatory molecules [2,78], and also substantial intake of anti-inflammatory food ingredients such as turmeric/curcumin [79,80].

The critical care units constitute a highly artificial environment and the burden of environment-induced physical and mental stress and subsequent status of systemic hyper-inflammation on the patient is unbearable. Patients treated under these conditions are in many aspects dys-functional, the whole microbiota is more or less gone and probiotic bacteria supplied will most often be killed already before they have reached the lower gastro-intestinal tract. The degree of artificiality seems to vary from hospital to hospital and from country to country, which might explain the great variation in outcome from studies undertaken in different regions and countries.

Probiotic treatment has never been given the chance as an alternative treatment, it has only this far been tried as a treatment complementary to all the other standard treatments. As discussed above, a series of auxiliary measures and particularly mechanical ventilation and treatment with various drugs, including antibiotics but also clinical nutrition solutions belong to those factors which promote super-inflammation and indirectly infection. Enteral nutrition formulas are also known to induce loss of intestinal barrier function, promotes bacterial translocation, and impair host immune defense, a phenomenon, observed in humans and further elucidated in animal studies.

# REFERENCES


1. Goldberg T, Cai W, Peppa M, Dardaine V, Baliga BS, Uribarri J, *et al.*: **Advanced glycoxidation end products in commonly consumed foods.** *J Am Diet Assoc* 2004, **104:**1287-1291.
2. Bengmark S: **Advanced glycation and lipoxidation end products - amplifiers of inflammation: the role of food.** *JPEN* 2007, **31:**430-440.
3. Hertoghe T: **The "multiple hormone deficiency" theory of aging: is human senescence caused mainly by multiple hormone deficiencies?** *Ann N Y Acad Sci* 2005, **1057:**448-465.
4. Moro JR, Iwata M, von Andriano UH: **Vitamin effects on the immune system: vitamins A and D take centre stage.** *Nat Rev Immunol* 2008, **8:**685-698.
5. McCarty MF: **Secondary hyperparathyroidism promotes the acute phase response - a rational for supplementing vitamin D in prevention of vascular events in elderly.**
6. *Medical Hypotheses* 2005, **64:**1022-1026.
7. Malekinejad H, Scherpenisse P, Bergwerff AA: **Naturally occurring estrogens in processed milk and in raw milk (from gestated cows).** *J Agric Food Chem* 2006, **54:**9785-9791.
8. Savoia C, Volpe M, Alonzo A, Rossi C, Rubattu S: **Natriuretic peptides and cardiovascular damage in the metabolic syndrome: molecular mechanisms and clinical implications.** *Clin Sci (Lond)* 2009, **118:**231-240.
9. Braly J, Hoggan R: *Dangerous grains*. Avery-Penguin Putnam Inc, NY; 2002:1-244.
10. Finegold SM, Sutter VL: **Fecal flora in different populations, with special reference to diet.** *Am J Clin Nutr* 1978, **31:**116-122.
11. Ahrné S, Nobaek S, Jeppsson B, Adlerberth I, Wold AE, Molin G: **The normal Lactobacillus flora in healthy human rectal and oral mucosa.** *J Appl Microbiol* 1998, **85:**88-94.
12. Topping DL, Fukushima M, Bird AR: **Resistant starch as a prebiotic and synbiotic: state of the art.** *Proc Nutr Soc* 2003, **62:**171-176.

13. Anderson JW: **Whole grain protect against atherosclerotic cardiovascular disease.**
14. *Proc Nutr Soc* 2003, **62:**135-142.
15. Dunjic BS, Svensson I, Axelsson J, *et al.*: **Is resistance to phospholipase important to gastric mucosal protective capacity of exogenous phosphatidylcholine?** *Eur J Gastroenterol Hepatol* 1994, **6:**593-598.
16. Dunjic BS, Svensson I, Axelsson J, Ar'Rajab A, Larsson K, Bengmark S: **Green banana protection of gastric mucosa against experimentally induced injuries in rats - a multicomponent mechanism?** *Scand J Gastroenterol* 1993, **28:**894-898.
17. Müller M, Lier D: **Fermentation of fructans by epiphytic lactic acid bacteria.** *J Appl Bacteriol* 1994, **76:**406-411.
18. Kaplan H, Hutkins RW: **Fermentation of fructooligosaccharides by lactic acid bacteria and Bifidobacteria.** *Appl Environ Microbiol* 2000, **66:**2682-2684.
19. Roma E, Adamidis D, Constantopoulos A, Messaritakis J: **Diet and chronic constipation in children: the role of fiber.** *J Pediatr Gastroenterol Nutr* 1999, **28:**169-174.
20. Moore N, Chao C, Yang LP, Storm H, Oliva-Hemker M, Saavedra JM: **Effects of fructo-oligosaccahride-supplemented infant cereal: a double-blind randomized trial.** *Br J Nutr* 2003, **90:**581-587.
21. Rushdi TA, Pichard C, Khater YH: **Control of diarrhea by fiber-enriched diet in ICU patients on enteral nutrition: a prospective randomized controlled trial.** *Clin Nutr* 2004, **23:**1344-1352.
22. Baqui AH, Black RE, El Arifeen S, Yunus M, Chakraborty J, Ahmed S, *et al.*: **Effect of zinc supplementation started during diarrhoea on morbidity and mortality in Bangladeshi children: community randomised trial.** *BMJ* 2002, **325:**1059.
23. Rabbani GH, Teka T, Zaman B, Majid N, Khatun M, Fuchs GJ: **Clinical studies in persistant diarrhea: dietary management with green banana or pectin in Bangladesh children.** *Gastroenterology* 2001, **121:**554-560.

24. Tucker KL, Hannan MT, Chen H, Cupples LA, Wilson PWF, Kiel DP: **Potassium, magnesium, and fruit and vegetable intakes are associated with greater bone mineral density in elderly men and women.** *Am J Clin Nutr* 1999, **69:**727-736.

25. Tucker KL, Chen H, Hannan MT, Cupples LA, Wilson PW, Felson D, *et al.*: **Bone mineral density and dietary patterns in older adults: the Framington osteoporosis study.** *Am J Clin Nutr* 2002, **76:**245-252.

26. Hylander B, Rössner S: **Effects of dietary fiber intake before meals on** WEIGHT LOSS AND **hunger in a weight-reducing club.** *Acta Med Scand* 1983, **213:**217-220.

27. Howarth NC, Saltzman E, McCrory MA, Greenberg AS, Dwyer J, Ausman L, *et al.*:**Fermentable and nonfermentable fiber supplements did not alter hunger, satiety or body weight in a pilot study in men and women consuming self-selected diet.**

28. *J Nutr* 2003, **133:**3141-3144.

29. Fernandez-Banares F, Hinojosa J, Sanchez-Lombrana JL, Navarro E, Martinez-Salmeron JF, Garcia-Puges A, *et al.*: **Randomized clinical trial of Plantago ovata seeds (dietary fiber) as compared with mesalamine in maintaining remission in ulcerative colitis. Spanish Group for the Study of Crohn's Disease and Ulcerative Colitis (GETECCU).** *Am J Gastroenterol* 1999, **427:**427-433.

30. Kanauchi O, Iwanaga T, Mitsuyama K: **Germinated barley foodstuff feeding. A novel neutraceutical strategy for ulcerative colitis.** *Digestion* 2001, **61:**S60-S67.

31. Wischmeyer PE, Riehm J, Singleton KD, Ren H, Musch MW, Kahana M, *et al.*: **Glutamine attenuates tumor necrosis factor-α release and enhances heat shock protein 72 in human peripheral blood mononuclear cells.** *Nutrition* 2003, **19:**1-6.

32. Hallert C, Björck I, Nyman M, Pousette A, Granno C, Svensson H: **Increasing fecal butyrate in ulcerative colitis patients by diet: controlled pilot study.** *Inflam Bowel Dis* 2003, **9:**116-121.

33. Lührs H, Gerke T, Müller JG, Melcher R, Schauber J, Boxberge F, *et al.*: **Butyrate inhibits NF-κB activation in lamina propria macrophages of patients with ulcerative colitis.** *Scand J Gastroenterol* 2002, **37:**458-466.

34. Welters CF, Heineman E, Thunnissens FB, van den Bogaard AE, Soeters PB, Baeten CG:**Effect of dietary inulin supplementation on inflammation of pouch mucosa in patients with an ileal pouch-anal anastomosis.** *Dis Colon Rectum* 2002, **45:**621-627.
35. Murch SH: **The immunologic basis for intestinal food allergy.** *Curr Opin Gastroenterol* 2000, **15:**552-557.
36. Bommelaer G, Dorval E, Denis PH, *et al.*: **Prevalence of irritable bowel syndrome in the French population according to Rome I criteria.** *Gastroenterol Clin Biol* 2002, **26:**11218-11223.
37. Wilhelm SM, Brubaker CM, Varcak EA, Kale-Pradhan PB: **Effectiveness of probiotics in the treatment of irritable bowel syndrome.** *Pharmacotherapy* 2008, **28:**496-505.
38. MacFarlane S, MacFarlane GT, Cummings JH: **Review article: prebiotics in the gastrointestinal tract.** *Aliment Pharmacol Ther* 2006, **24:**701-714.
39. Spiller R: **Review article: probiotics and prebiotics in irritable bowel syndrome (IBS).** *Aliment Pharmacol Ther* 2008, **28:**385-396.
40. Seifert S, Watzl B: **Prebiotics and the Immune System: Review of Experimental and Human Data.** In *Handbook in prebiotics.* Edited by Gibson G, Roberfroid M. CRC Press, Boca Raton; 2008.
41. Lindsay JO, Whelan K, Stagg AJ, *et al.*: **Clinical, microbiological, and immunological effects disease of fructo-oligosaccharide in patients with Crohn's disease.** *Gut* 2006, **55:**348-355.
42. Huertas-Ceballos A, Logan S, Bennett C, Macarthur C: **Dietary interventions for recurrent abdominal pain (RAP) and irritable bowel syndrome (IBS) in childhood.** *Cochrane Database Syst Rev* 2008, (1)**:**CD003019.
43. Parisi GC, Zilli M, Miani MP, *et al.*: **High-fiber diet supplementation in patients with irritable bowel syndrome (IBS): a multicenter, randomized, open trial comparison between wheat bran diet and partially hydrolyzed guar gum (PHGG).** *Dig Dis Sci* 2002, **47:**1697-1704.
44. de Felippe Junior J, da Rocha e Silva Junior M, Maciel FM, Soares Ade M, Mendes NF:**Infection prevention in patients with severe multiple trauma with the immunomodulator beta 1–3 polyglucose (glucan).** *Surg Gynecol Obstet* 1993, **177:**383-388.

45. Caparros T, Lopez J, Grau T: **Early enteral nutrition in critically ill patients with a high- protein diet enriched with arginine, fiber, and antioxidants compared with a standard high-protein diet.** *The effect on nosocomial infections and outcome. JPEN J Parenter Enteral Nutr* 2001,**25:**299-308.

46. Suzuki C, Kimoto-Nira H, Kobayashi M, Nomura M, Sasaki K, Mizumachi K:**Immunomodulatory and cytotoxic effects of various Lactococcus strains on the murine macrophage cell line J774.1.** *Int J Food Microbiol* 2008, **123:**159-165.

47. Peuranen S, Tiihonen K, Apajalkathi Kettunen A, Saarinen M, Rautonen N: **Combination of polydextrose and lactolol affects microbial ecosystem and immune responses in rat gastrintestinal tract.** *Br J Nutr* 2004, **91:**905-914.

48. Timmermann HM, Koning CJM, Mulder L, Rombouts FM, Beynen AC: **Monostrain, multistrain and multispecies probiotics - a comparison of functionality and efficacy.**

49. *Int J Food Microbiol* 2004, **96:**219-233.

50. Johansson ML, Molin G, Jeppsson B, Nobaek S, Ahrne S, Bengmark S: **Administration of different lactobacillus strains in fermented oatmeal soup: in vivo colonization of human intestinal mucosa and effect on the indigenous flora.** *Appl Environ Microbiol* 1993, **59:**15-20.

51. Ljungh Å, Lan J-G, Yamagisawa N: **Isolation, selection and characteristics of Lactobacillus paracasei ssp paracasei isolate F19.** *Microb Ecol Health Dis Suppl* 2002, **3:**4-6.

52. Kruzewska K, Lan J, Lorca G, Yanagisawa N, Marklinder I, Ljungh Å: **Selection of lactic acid bacteria as probiotic strains by in vitro tests.** *Microecol Ther* 2002, **29:**37-51. The proceedings of the XVI international Congress on Microbial Ecology and Disease, held Noordwijkerhout, The Netherlands, Oct 2001

53. Fujiwara D, Inoue S, Wakabayashi H, Fujii T: **The anti-allergic effects of lactic acid bacteria are strain dependent and mediated by effects on both Th1/Th2 cytokine expression and balance.** *Int Arch Allergy Immunol* 2004, **135:**205-215.

54. von der Weid T, Bulliard C, Schiffrin EJ: **Induction by a lactic acid bacterium of a population of CD4(+) T cells with low proliferative capacity that produce transforming growth factor**

**beta and interleukin-10.** *Clin Diagn Lab Immunol* 2001, **8:**695-701.

55. Ibnou-Zekri N, Blum S, Schiffrin EJ, von der Weid T: **Divergent patterns of colonization and immune response elicited from two intestinal Lactobacillus strains that display similar properties in vitro.** *Infect Immun* 2003, **71:**428-436.
56. Nagler-Anderson C: **Tolerance and immunity in the intestinal immune system.** *Crit Rev Immunol* 2000, **20:**103-120.
57. Prioult G, Fliss I, Pecquet S: **Effect of probiotic bacteria on induction and maintenance of oral tolerance to beta-lactoglobulin in gnotobiotic mice** *Clin Diagn Lab Immunol* 2003, **10:**787-792.
58. Naaber P, Smidt I, Stsepetova J, Brilene T, Annuk H, Mikelsaar M: **Inhibition of lostridium difficile strains by intestinal Lactobacillus species.** *J Med Microbiol* 2004, **53:**551-554.
59. Verdú EF, Bercík P, Bergonzelli GE, Huang XX, Blennerhasset P, Rochat F, Fiaux M, Mansourian R, Corthésy-Theulaz I, Collins SM: **Lactobacillus paracasei normalizes muscle hypercontractility in a murine model of postinfective gut dysfunction.** *Gastroenterology* 2004, **127:**826-837.
60. Eutamene H, Lamine F, Chabo C, Theodorou V, Rochat F, Bergonzelli GE, Corthésy-Theulaz I, Fioramonti J, Bueno L: **Synergy between Lactobacillus paracasei and its bacterial products to counteract stress-induced gut permeability and sensitivity increase in rats.** *J Nutr* 2007, **137:**1901-1907.
61. Oláh A, Belágyi T, Pótó L, Romics L, Bengmark S: **Synbiotic control of inflammation and infection in severe acute pancreatitis, a randomized double blind study.** *Hepatogastroenterology* 2007, **54:**36-41.
62. Spindler-Vesel A, Bengmark S, Vovk I, Cerovic O, Kompan L: **Synbiotics, prebiotics, glutamine, or peptide in early enteral nutrition: a randomized study in trauma patients.** *JPEN J Parenter Enteral Nutr* 2007, **31:**119-126.
63. Kotzampassi K, Giamerellos-Bourboulis EJ, Voudouris A, Kazamias KP, Eleftheriadis E:**Benefits of Synbiotic 2000 Forte in critically ill trauma patients -early results of a randomized controlled trial.** *World J Surg* 2006, **30:**1848-1855.

64. Giamarellos-Bourboulis EJ, Bengmark S, Kanellakopoulou K, Kotzampassi K: **Pro- and synbiotics to control inflammation and infection in patients with multiple injuries.** *J Trauma* 2009, **67:**815-821.

65. Rayes Rayes N, Seehofer D, Theruvath T, Mogl M, Wilke C, Schiller RA, *et al.*: **Supply of pre- and probiotics reduces bacterial infection rates after pylorus-preserving pancreatoduodenectomy - a randomized, double-blind trial.** *Ann Surg* 2007, **246:**36-41.

66. Liu Q, Zhong Ping Duan, Da Kang Ha, Bengmark S, Kurtovic J, Riordan SM: **Synbiotic modulation of gut flora: effect on minimal hepatic encephalopathy in patients with liver cirrhosis.** *Hepatology* 2004, **39:**1441-1449.

67. Riordan SM, Skinner NA, McIver CJ, Lio Q, Bengmark S, Bihari D, *et al.*: **Synbiotic-associated improvement in liver function in cirrhotic patients: Relation to changes in circulating cytokine messenger RNA and protein levels.** *Microb Ecol Health Dis* 2007, **19:**7-16.

68. Rayes N, Seehofer D, Theruvath T, Schiller RA, Langrehr JM, Jonas S, *et al.*: **Combined perioperative enteral supply of bioactive pre- and probiotics abolishes postoperative bacterial infections in human liver transplantation - a randomised, double blind clinical trial.** *Am J Transplant* 2005, **5:**125-130.

69. Pathmakanthan S, Walsh M, Bengmark S, *et al.*: **Efficacy and Tolerability treating acute distal ulcerative colitis with synbiotic enema's: A Pilot Trial. Abstract.** *Gut* 2002, **51**(suppl 1III)**:**A307.

70. Rutgeerts P, D'Haens G, Baert F, Van Assche M, Noman I, Aerden S, *et al.*:**Randomized placebo controlled trial of pro-and prebiotics (synbiotics cocktail) for maintenance of infliximab induced remission of luminal Crohn's disease (CD).**

71. *Gastroenterology* 2004, **126:**A-467 T1310

72. Chermesh I, Tamir A, Reshef R, Chowers Y, Suissa A, Katz D, *et al.*: **Failure of Synbiotic 2000 to prevent postoperative recurrence of Crohn's disease.** *Dig Dis Sci* 2007, **52:**385-389.

73. Gomersall CD, Joynt GM, Leung P, Tan P, Bengmark S: *Does routine administration of probiotics improve outcome of critically ill patients?*. Abstract, ANZCA ASM; 2006.

74. Knight D, Girling K, Banks A, Snape S, Weston W, Bengmark S: **The effect of enteral synbiotics on the incidence of ventilator associated pneumonia in mechanically ventilated critically ill patients Abstract.** *Brit J Anaesth* 2004, **92:**307P-308P.

75. Stampfer MJ, Hu FB, Manson JE, *et al.*: **Primary prevention of coronary heart disease in women through diet and lifestyle.** *N Engl J Med* 2000, **343:**16-22.

76. Hu FB, Manson JE, Stampfer MJ, *et al.*: **Diet, lifestyle and the risk of type 2** DIABETES MELLITUS **in women.** *N Engl J Med* 2001, **345:**790-797.

77. Platz EA, Willett WC, Colditz GA, Rimm EB, Spiegelman D, Giovannucci E: **Proportion of colon cancer risk that might be preventable in a cohort of middle-aged US men.** *Cancer Causes Control* 2000, **11:**579-588.

78. Haller D, Antoine JM, Bengmark S, Enck P, Rijkers GT, Lenoir-Wijnkoop I: **Guidance for Substantiating the Evidence for Beneficial Effects of Probiotics: Probiotics in Chronic Inflammatory Bowel Disease and the Functional Disorder Irritable Bowel Syndrome.** *J Nutr* 2010

79. Wang TT, Dabbas B, Laperriere D, Bitton AJ, Soualhine H, Tavera-Mendoza LE, Dionne S, Servant MJ, Bitton A, Seidman EG, Mader S, Behr MA, White JH: **Direct and indirect induction by 1,25-dihydroxyvitamin D3 of the NOD2/CARD15-defensin beta2 innate immune pathway defective in Crohn disease.** *J Biol Chem* 2010, **285:**2227-2231.

80. Pachikian BD, Neyrinck AM, Deldicque L, De Backer FC, Catry E, Dewulf EM, Sohet FM, Bindels LB, Everard A, Francaux M, Guiot Y, Cani PD, Delzenne NM: **Changes in intestinal Bifidobacteria levels are associated with the inflammatory response in magnesium-deficient mice.** *J Nutr* 2010, :.

81. Bengmark S: **Nutritional modulation of acute and "chronic" phase response.**

82. *Nutrition* 2001, **17:**489-495.

83. Bengmark S: **Acute and "chronic" phase response - a mother of disease.** *Clin Nutr* 2004, **23:**1256-1266.

84. Bengmark S: **AGE, ALE RAGE and disease - a foods perspective.** In *Handbook of Prebiotics and Probiotics Ingredients: Health*

*Benefits and Food Applications*. Edited by Cho Susan S. Terry Finocchiaro CRC Press, Taylor and Francis Group, Boca Raton; 2009.

85. Bengmark S: **Curcumin an atoxic antioxidant and natural NF-κB, COX-2, LOX and iNOS inhibitor - a shield against acute and chronic diseases.** *JPEN J Parenter Enteral Nutr* 2006, **30:**45-51.
86. Bengmark S: **Control of systemic inflammation and chronic diseases - the use of turmeric and curcuminoids.** In *Nutrigenomics and Proteogenomics in Health and Disease: Impact of Food Factors-Gene Interactions*. Edited by Mine Y, Miyashita K, Shahidi F. John Wiley and Son, Oxford; 2009.

Chapter 7

# A Review on Integrated Agro-Technology of Vegetables

T. S. Workneh[1] and G. Osthoff[2]

[1]School of Bioresources Engineering and Environmental Hydrology, Faculty of Engineering, University of Kwa-Zulu Natal, SCOTTSVILLE, 3209, South Africa

[2]Department of Food Science, Facult y of Agriculture, The University of Free State, Bloemfontein 3900, South Africa

## ABSTRACT

The aim of this review was threefold: First, to explore the effect of different preharvest treatments on postharvest quality of fruits and vegetables. Second, the principles of biological, chemical and biochemical changes in fruits and vegetables during development, maturation, ripening and storage were reviewed. Third postharvest handling and factors affecting quality of fruits and vegetables were examined. These include disinfecting, packaging and storage temperature. Pre- and postharvest treatments were found to have an effect on postharvest quality of fruit and vegetables, suggesting that postharvest quality of produce subjected to preharvest treatments should be assessed from a quality improvement, maintenance and consumer safety point of view. Literature recommends an integrated agro-technology approach towards improving quality at harvest and maintenance of qualities of fruits and vegetables.

# INTRODUCTION

An understanding of the changes of fruits and vegetables during storage entails more than just knowledge of packaging methods. Preharvest as well as postharvest physiological properties have to be understood. Therefore, in this chapter, the survey on the effect of preharvest treatments on postharvest quality of fresh commodities is presented first. This is followed by the survey of postharvest handling and storage including pre-packaging treatments and storage methods available for fresh commodities. Since tomatoes and carrots were selected for this study, the survey will mainly concentrate on these vegetables.

# DEVELOPMENT PHYSIOLOGY OF FRUIT AND VEGETABLES

## Development Physiology of Tomatoes

Tomato physiology begins with fertilisation of the ovules of the blossom (Salunkhe et al., 1991). Hormone production by the developing seeds and young ovary walls are highly responsible for growth. The period from the end of flowering to and including the ripening stage, during which chemical changes take place and new tissue is formed and brought to morphological completion, is known as the development period (Salunkhe et al., 1991). It includes stages of permutation, physiological maturation, ripening, and senescence (Kader et al., 1985). These stages are followed by continued changes in the chemical composition, which in turn is governed by a range of enzyme activities (Kader et al., 1985).

Tomatoes are commercially mature at a fully developed fruit stage. Endogenous ethylene is present in measureable quantities during the entire development of the tomato fruit (Lyons and Pratt, 1964). It has been reported that soluble peroxidase activity increases dramatically during the early stages of tomato fruit development, reaching a maximum at the green mature stage (Thomas et al., 1981). The soluble peroxidase activity remained higher during the breaker and light pink stages, but decreases in the later stages of development (Thomas et al., 1981). Simultaneously IAA (indoleacetic acid) oxidase activity in the soluble fraction followed a parallel pattern to the peroxidase activity (Frenkel, 1972; Thomas et al., 1981). Frenkel (1972) reported that the induction of the major isozyme component of peroxidase and IAA oxidase was enhanced as the tomato fruit ripens. The amount of auxin protectors also increased as the fruit developed (Thomas et al., 1981).

A relationship between nicotinamide adenine dinucleotide phosphate ($NADP^+$)-malic enzyme and organic acid levels exists in tomatoes from flowering through to ripening, and both increase during develop-ment, reaching maximum levels at the green mature stage (Knee and Finger, 1992). However, a decline in malate concentration is followed by a decrease in NADP-malic enzyme activity and citrate concentration. Their data reported demonstrated that, it is possible that an enzyme is involved in cytoplasmic pH regulation. The sugar content of tomato fruits increases progressively throughout maturation and ripening, with a pronounced increase with the appearance of yellow pigmentation (Winsor et al., 1962a, b, in Hobson and Davies, 1971). The starch content of tomatoes also increases with maturation, reaching a maximum at 8 weeks after fertilisation, but is not detected in young fruit up to 10 days. Results on the ascorbic acid content of tomatoes during the development and ripening seemed inconsistent. However, recent studies have indicated an increase in ascorbic acid contents of tomato fruit during maturation, with either a continuing increase or a slight decrease during the final stages of ripening (Dalal et al., 1965; Mohammed et al., 1999). The malic acid concentration decreases as tomatoes ripen, while citric acid increases up to the greenyellow stage and then generally decreases (Hobson and Davies, 1971).

## Development Physiology of Carrots

Phan et al. (1973) gathered comprehensive information on the carrot roots during growth up to harvest. The main constituents of carrot roots are soluble carbohydrates comprising of non-reducing sugars, mostly sucrose, and reducing sugars, mostly glucose. Their data showed that there was active biosynthesis of carbohydrates, mainly sucrose, and carotenes, such as b-carotene (Phan et al., 1973). The sugars and carotene contents of carrot roots consistently increase during the 3 months after seeding and reach their maxima at the end of 3 months. After 3 months of the development period, the sugar content of both groups of substances remains almost constant. The total soluble carbohydrate concentration increases rapidly a few days after the initiation of an entire ring of cambium in the carrot roots (Hole and McKee, 1988). Their data concerning the relationship between enzyme activity and carbohydrates during the development of carrot roots revealed the existence of little correlation.

The amount of organic and amino acids increases slowly with age during root development and these components are present in rather low concentrations (Phan et al., 1973). However, pyruvic acid occurs in high amounts in growing carrot roots up to 3 months, which is followed by isocitric acid and malic acid (Phan et al., 1973). These results indicated that a 3-month growth period for carrots could be sufficient before harvest, after which there is no more increase in chemical components responsible for good quality characteristics. However, this "biochemical maturity" is reached while carrots are still growing in diameter and can therefore not be taken as a criterion for determining the harvest date (Phan et al., 1973).

# PREHARVEST TREATMENTS OF VEGETABLES

## Effect of Preharvest Factors on Storage Quality of Vegetables

Preharvest treatments of fruits and vegetables are primarily aimed at increasing yields, while postharvest storage performance is normally neglected. Several research results were reported on methods to increase harvest yield and qualities of fruits and vegetables (Mitchell et al., 1997). Most research work has been targeted on cultural practices, such as rootstock/plant age, soil management, nutrition, training and pruning practices, crop loads, product size, and growth regulators (Rosenfeld, 1999). Bramlage (1993) in Watkins and Pritts (2001) highlighted the almost overwhelming number of preharvest variables that contribute to the variety of postharvest responses of the crops. Watkins and Pritts (2001) hypothesise that the diversified postharvest responses of fruits and vegetables during storage are in part due to preharvest cultural practice. A literature review has shown that the major factors affecting yield and quality of vegetables are cultivars, soil plant systems and fertiliser practices, and the environmental factors such as temperature, relative humidity, light intensity and rainfall during production (Rosenfield, 1999).

## Soil Plant System and Fertiliser Practice

From a literature review on the effect of cultural practice on quality of vegetables with emphasis on tomatoes, carrots and lettuce (Rosenfeld, 1999), it was deduced that in general, the objectives of optimal fertilisation strategy were maximization of yield, the maximization of fruit and vegetable quality, minimization of environmental pollution caused by fertilisers and the minimisation of fertiliser expenses. The yield of potatoes, soybeans, cabbages, carrots, onions, cucumbers, strawberries, and eggplants oscillated because of the different soil and climatic conditions (Data given by Polus, in Schnabl et al., 2001). These could also suggest that there could be changes in postharvest quality based on different soil and climatic conditions.

Postharvest quality parameters of tomatoes and carrots vary with the fertilisation practice during production. Even under unfavourable climatic conditions, application of phosphorous-potassium (PK) fertiliser could be used to increase carotene content in carrots (Evers, 1989a). Photosynthetic products like sugar are slightly affected with fertilisation (Evers, 1989a). However, Evers (1989a) showed that seasonal variations and genetic variations are often larger than variations caused by soil and fertiliser practices.

Qualities of carrots are reduced as a result of increased use of mineral fertiliser, but are not affected by measured fertilizers (Lieblein, 1993) whereas increased use of composted manure had no effect on quality of carrots. The postharvest response of carrots is often dependent on the level of fertiliser applied during the growth period (Petrichenko et al., 1996).

Vitamin B concentrations are higher in plants grown with organic fertilisers as compared with plants grown with inorganic fertilisers (Mozafar, 1994). Nitrogen fertilisers, especially at high rate, seem to decrease the concentration of vitamin C in different fruits and vegetables, among them are potatoes, tomatoes and citrus fruits (Mozafar, 1994).

Positive and negative effects of preharvest treatments on tomatoes, either to increase yield or improve nutritional quality, have been reported (Gao et al., 1996; Carmer et al., 2001). Carmer et al. (2001) sup-plied tomato plants with either low electric conductivity (EC = 0.25 $Sm^{-1}$) nutrient solutions or with nutrient solutions supplied with 55 mM NaCl to generate at high electric conductivity (0.75 $m^{-1}$). Their results showed that high electric conductivity increased the total soluble solids (TSS) by ca. 18% and titratable acids by ca. 32% relative to low electric conductivity treatments. The report also indicted that after storage of 2 weeks at 15°C, fruits of high electric conductivity treated plants were 12% less firm than those of low electric conductivity plants, and no difference in TSS or acidity were found.

Salunkhe et al. (1971) has shown the effect of Telone (1,3-dichloropropene and other chlorinated hydropropane) and Nemagon (1,2-dibromo-3-chloropropane) on essential nutritive components and the respiratory rates of carrots and sweet corn seeds. The respiration of carrots treated with Telone and Nemagon was below 94 ml $O_2$ $h^{-1}$ $g^{-1}$ and 82.8 ml $O_2$ $h^{-1}$ $g^{-1}$, respectively, while that of untreated carrots was 108.8 ml $O_2$ $h^{-1}$ $g^{-1}$. This result clearly showed that preharvest treatments could affect the physiology of carrots at harvest. A significant increase in the content of total carotene, _-carotene and total sugars, was observed, with a simultaneous decrease in respi- Workneh and Osthoff 9309 ratory rates in carrots (Salunkhe et al., 1971). The low respiration rates indicated that the metabolic activities of the treated carrots were low, which lead to increased shelf life without quality deterioration. Pre-planting soil fumigation with Telone and Nemagon also resulted in increased carotene, b-carotene and total sugars, and decreased the respiratory rates in carrot roots (Singh et al., 1970).

Most researchers reported that there are either positive or negative effects of any type of preharvest practices on quality of carrots and tomatoes, especially at harvest (Watkins and Pritts, 2001; Tittonell et al., 2001; Ozeker et al., 2001; Sen et al. 2001). Salt concentrations of nutrient solutions were shown to affect quality of celery more than yield (Pardossi et al., 1999), and preharvest Ethephon (2-chloroethylphosphonic acid) spray directly onto pepino fruits advanced colour changes (Maroto et al., 1995; Lopez et al., 2000). Nutritional treatments had a positive effect in reducing the peel disorder of fruits under commercial conditions (Zilkah et al., 2001). More research is needed on the storage of fresh produce subjected to various preharvest practices. It is also recommended that after each preharvest practice a study on the quality of vegetables and their storability needs to be conducted in order to secure favourable storage conditions for these products and maintain freshness and nutritional quality.

## Effect of Environmental Factors on Vegetable Quality

The effect of climate on quality of vegetables is normally mostly higher than the effect of fertiliser on photosynthetic products such as sugar (Rosenfeld, 1999). Temperature is the major environmental factor affecting quality of vegetables. Vegetables like carrots and tomatoes respond to various levels of temperature. Carrots grown at high temperature were shown to have a higher total sugar content, whereas those grown at low temperature were sweeter, specifically due to a higher sucrose content (Rosenfeld, 1999). Apple watercore, ethylene evolution, flesh firmness, membrane permeability and sorbitol levels were shown to be affected by preharvest fruit temperature (Yamada and Kobayashi, 1999).

Hao and Papadopoulos (1999) have shown that supplemental lighting during growth of cucumber increases biomass allocation to fruits, fruit dry matter content and skin chlorophyll content. Leonardi et al. (2000) grew tomato plants in two glasshouse compartments under two vapour pressure deficit (vpd) levels, showing that fruit growth and transpiration rates greatly varied during daylight hours, which has enhanced under high vpd conditions. As a result a significant reduction in fruit weight and in fruit water content, and an increase in soluble solids was found. Environmental vapour pressure increases can therefore affect not only growth but also quality of tomato fruits. It was shown that air humidity has effects on growth, flowering, and finally on keeping quality of some greenhouse species (Mortensen, 2000).

## Water Management

Sørensen et al. (1997) showed the effect of drought stress on carrot quality. Glucose, fructose and sucrose concentrations in carrots exposed to drought stress at different growth stages were not affected in any consistent manner. The results also indicated that the concentration of dry matter was low when drought was induced at an early growth stage in coarse and sandy soil. Averaging the effect of drought periods and cultivars, drought stress was observed to increase the concentration of sucrose in taproots from the coarse, sandy soil. It was also shown that drought stress increased storage losses due to the development of disease (Sørensen et al., 1997).

Research on deficit-irrigating micro-irrigated tomatoes showed that occurrence of early blight disease (caused by the fungus *Alternaria solani*) was increased by 50%, while blossom end rot incidence was five times more severe compared with full irrigation (Obreza et al., 1996). This result indicated that deficit irrigation could cause substantial economic loss of tomatoes through decreased crop marketable quality.

Dodds et al. (1996) studied the influence of water table depth and found that 0.6 m depth gave the best yield and largest fruit size, however, a higher incidence of catfacing, cracking, sunscald and loss of firmness of tomatoes were found. A balance between yield and quality at a water table depth between 0.6 and 0.8 m was recommended for tomato production on sandy loam soils.

On the other hand, irrigation deficit in the first growth period of tomato reduced the number of flowers leading to a decrease in the number of fruits and in the marketable yield (Colla et al., 1999). The soil moisture deficit resulted in increased soluble solids and acidity of the fruit. However, reducing irrigation by 25% before fruit onset and by as much as 50% in the fruit development and ripening stages resulted in no significant decrease of soluble solid yield.

## Hormone Treatment

Hormones are essential for plant growth and development. The quality of hormone present and tissue sensitivity to hormones has an effect on plant physiology. Plant hormones are responsible for cell elongation, cell division, inhibition of senescence, abscission of leaves and fruits, dormancy induction of buds and seeds, promotion of senescence, epinasty and fruit ripening. The preharvest as well as postharvest physiological processes in fruits and vegetables are responsible for changes in composition and qualityof these horticultural crops and depend on the plant species. The major classes of plant hormones include auxins, gibberelins, cytokinins, abscisic acid, brassinosteroids and ethylene (Mauseth, 1991; Raven et al., 1992; Salisbury and Ross, 1992; Davies, 1995).

Gibberellins are known to stimulate the physiological processes such as stem elongation, bolting, breaking seed dormancy, enzyme production, induce maleness in dioecious flowers, cause parthenocarpic (seedless) fruit development and can delay senescence in leaves and citrus fruits (Mauseth, 1991; Raven et al., 1992; Salisburg and Ross, 1992; Davies, 1995). Cytokinin stimulates cell division, morphogenesis in tissue, the growth of lateral buds, release of apical dominance, leaf expansion resulting from cell enlargement, enhances stomatal opening in some species and promotes the conversion of etioplasts into chloroplasts via stimulation of chlorophyll synthesis (Mauseth, 1991; Raven et al., 1992; Salisbury and Ross, 1992; Davies, 1995). The functions of abscisic acid are to stimulate the closure of stomata, inhibition of shoot growth, induction of seeds to synthesise storage proteins, induction of gene transcription, especially for proteinase inhibitors in response to wounding, which may explain an apparent role in pathogen defence (Mauseth, 1991; Raven et al., 1992; Salisbury and Ross, 1992; Davies, 1995). Ethylene has been the most studied plant hormone in relation to fruit ripening and postharvest storage. Some of the functions of ethylene are to stimulate the release of dormancy, shoot and root growth, and differentiation of adventitious root formation, leaf and fruit abscission, bromiliad flower induction, induction of femaleness in dioecious flowers, flower opening, flower and leaf senescence and fruit ripening (Mauseth, 1991; Raven et al., 1992; Salisbury and Ross, 1992; Davies, 1995).

The other growth regulating compounds are brassinosteroids, salicylates, jasmonates and polyamines. An abundance of research has been devoted on the chemistry and physiology of natural growth regulators resulting in the recognition of brassinosteroids as a new class of phytohormones (Schnabl et al., 2001). Some of the effects of brassinosteroids include stimulation of stem elongation, inhibition of root growth and development and promotion of ethylene biosynthesis and epinasty. Recently, the brassinosteroids has gained a broad spectrum of application and extremely low toxicity and mutagenicity have received increasing attention (Schnabl et al., 2001). Previous work showed that brassinosteroids are plant growth promoting regulators, which are effective in cell elongation and division, source/ sink metabolism, chlorophyll synthesis and reproductive and vascular development (Clouse, 1996; Mandalla, 1988). Sasse et al. (1995) and Takatsuto et al. (1996) reported that they enhance nutrient contents, improve shape and taste of fruits, have beneficial effects on germination, growth and seed quality. Much research has been conducted on the potential performance and investigations of the potential applications of brassinosteroids in agriculture (Schnabl et al., 2001).

## Communication Catalization (ComCat®) Treatment

Recently, a hormone containing treatment has been introduced as an alternative agricultural input to the use of chemicals to increase production of vegetables and other crops. ComCat® is a natural biocatalyst, which is extracted from seeds of plants and mainly consists of amino acids, gibberellin, kitenins, auxin (indole-3-acitic acid), brassinosteroids, natural metabolites, pathogen-related PR-proteins with defence reactions, terpenoids, flavonoids, vitamins, inhibitors, other signal molecules, biocatalysts and cofactors. The yields of ComCat® treated vegetables were shown to be increased for cabbage (8%), tomato (16 to 19%), potato (9 to 19%), soybeans (26 to 30%), eggplants (37%), cucumbers (25 to 32%), carrots (32%), onions (49%) and strawberries (50%), compared to control vegetables and fruits (Schnabl et al., 2001). These vegetables were also shown to have better root development, improved resistance induction, less chance of deficiencies with fertiliser, higher resistance to pathogens prior to harvest and they seem to have a slightly better resistance to environmental stress, and increased protein content (Hüster, 2001; Schnabl et al., 2001).

A reduction of plant disease symptoms of up to 45% in comparison to untreated control plants has been found. This is the result of induction of the PR-proteins (pathogenesis-related proteins) namely peroxidase, chitinase and 1-3 gluconase. These enzymes protect cell walls and prevent infection by fungi (Hüster, 2001). Advantages of ComCat® treatment are that only low doses are necessary to show measurable effects of these brassinosteroids containing plant extract in crop plants, their environmental safety and the possibility of reducing the amount of pesticides needed (Schnabl et al., 2001). Apart from studies on yield increase, no data is yet available on the effect of preharvest ComCat® treatment on quality of fruits and vegetables at harvest, as well as during storage.ComCat® was approved by, and registered with the Federal German Biological Centre of Agriculture and Forestry (BBA), Institute for Integrated Plant protecttion, as a harmless plant strengthening substance of plant origin. It is also licensed for use in ecological farming, according to the EU-regulation 2092/91.

As discussed, the effects of ComCat® include (a) serving the plant as a general strengthening agent for the organic development, (b) inducing resistance through activating plant own defence mechanisms against pathogens, and biotic and abiotical stress factors, (c) improve root development, (d) increase yield in agricultural cash crops as well as horticultural crops and (e) increase protein content.

# POSTHARVEST PHYSIOLOGY OF FRUIT AND VEGETABLES

## Respiration

Respiration is defined as a process by which stored Workneh and Osthoff 9311 organic materials (carbohydrates, proteins, and fats) are broken down into simple end products with a release of energy. In the process, $O_2$ is used and $CO_2$ is liberated. Vegetables continue to respire after harvest and during storage. Each type of vegetable and cultivars requires a specific range of $CO_2$ and $O_2$ concentration levels for safe storage without the occurrence of physiological disorder. The level of physiological activity and potential storage life can be indexed by the rate of respiration. Respiration is one of the basic physiological factors, which speeds up ripening of fresh commodities and is directly related to maturation, handling, and ultimately to the shelf life (Ryall and Lipton, 1979; Ryall and Pentzer, 1982). Generally, the loss of freshness of perishable commodities depends on the rate of respiration. An increase in respiration rate hastens senescence, reduces food value for consumers, and increases the loss of flavour and sellable dry weight.

Stored intact fruits and vegetables face desiccation and chilling injury after harvest and during storage. Due to wounding stress, as a result of chilling or mechanical injuries, respiration rate and overall metabolic activities usually increase. The main physiological manifestation of metabolic activities include increased respiration rate and, in some cases, ethylene production (Rosen and Kader, 1989).

The index of physiological activity and potential shelf life have a direct relationship with respiration of fruits and vegetables. Since sugars in fruits and vegetables play a role in the respiration process, the quantity of sugars in fruits and vegetables available for respiration is the dominant factor for the shelf life of these commodities at a given temperature (Paez and Hultin, 1972). For normal respiration, removal of respiratory $CO_2$ needs more emphasis than supply of $O_2$, because some fruits and vegetables are highly sensitive and could be suffocated with a high level of $CO_2$ (Duckworth, 1966; Kader et al., 1985). Removal of respiratory heat requires attention because it increases the product temperature and surrounding air temperature, which in turn is responsible for increasing respiration and causes acceleration of substrate utilization, predominantly sugars (Ryall and Lipton, 1979; Ryall and Pentzer, 1982). The rate of respiration depends on the quantity of available $O_2$ as well as the storage temperature. A decrease in rate of respiration increases the shelf life of fruits and vegetables. In order to achieve long storage life of fruits and vegetables, the rate of respiration should be reduced through decreasing the $O_2$ level, slightly increasing the $CO_2$ level, and decreasing the storage temperature, which includes removal of respiratory heat (Duckworth, 1966; Kader et al., 1985). The optimum gas composition is the range of $O_2$ and $CO_2$ levels that would minimise physiological disorder, reduce respiration rate and reduce ethylene production during storage. The limit of tolerance to low $O_2$ and high $CO_2$ levels depends on several parameters including temperature, physiological conditions, maturity, and previous treatment.

## Temperature Quotient of Respiration

Temperature highly affects the metabolic activities of fruits and vegetables. Relatively higher temperatures increase the rate of respiration and ethylene production during storage. The rate of chemical reactions in fruits and vegetables is also controlled by temperature. It was reported that theoretically, the rate of respiration doubles for each 10°C increase in temperature. Depending on the maturity and anatomical structure of the fruits or vegetables, the temperature quotient of respiration may be more than double (Ryall an Lipton, 1979; Ryall and Pentzer, 1982; Sargent et al., 1991). These researchers showed that the respiration rate of topped carrots increased by 79% at 25 to 27°C, when compared to topped carrots at 0°C. Similarly, the rate of respiration (rate of $CO_2$) of bunched carrots increased by about 71% at 25 to 27°C, compared to those at 0°C (Hardenburg et al., 1986), and the respiration rate of mature green tomatoes increased by 100% when stored at 25 to 27°C compared to storage at 0°C (Hardenburg et al., 1986). Preharvest treatments on a farm, or in an orchard, affects postharvest physiology of fruits and vegetables, such as the respiration rates. Physiology of fruits and vegetables begins at the time of blossoming or bud formation and is affected by preharvest factors such as fertilisation, variety, and irrigation, and by environmental factors such as sunlight duration and quality, temperature, humidity etc., as well as preharvest spray of hormones and growth regulators (Ryall and Lipton, 1979; Ryall and Pentzer, 1982; Schnabl et al., 2001). These treatments could possibly have positive or negative effects on the postharvest quality and shelf life of fruits and vegetables, indicating the importance of further integrated research on pre-and postharvest physiology when implementing new preharvest treatment agents or methods.

## Ethylene Production and Effects

Ethylene advances the onset of an irreversible rise in respiration rate in climacteric fruit and increases the ripening process. The effect of low $O_2$ and high $CO_2$ levels on the production of ethylene adds to the nature of the ethylene production or inhibition process. Fruits and vegetables are classified into five groups according to ethylene production rates within the ranges of 0.1 ml ethylene $kg^{-1}h^{-1}$ at 20°C to

100.0 ml ethylene $kg^{-1}$ $h^{-1}$ at the same temperature (Kader et al., 1985). In general, the ethylene production rates of tomatoes range from 1.0 - 10 ml $kg^{-1}h^{-1}$ at 20°C, which classifies them in the moderate class according to ethylene production rate (Ryall and Lipton, 1979; Kader et al., 1985). Tomato is one of the few vegetables to which a known phytohormone, ethylene, is applied commercially to influence the rate of ripening. Ethylene plays a significant role in the physiological and biochemical changes that occur with the climacteric onset. Lyons and Pratt (1964) reported that endogenous ethylene was present in measurable quantities during the entire growth phase of the tomato fruit. The concentration of ethylene increased 10-fold when fruit growth reached 70 to 93% of its total fruit growth (Lyons and Pratt, 1964).

It was also shown that the concentration of ethylene increased 400 times that of the average measured during growth, at the climacteric onset of ethylene production and onset of ripening. External introduction of ethylene to tomatoes at all stages of development induces ripening and climacteric onset along with phenotypic changes common to normal ripening, such as red color development, fruit softening, and characteristic flavour (McGlasson, 1978 and 1985). However, the acceptable edible quality of tomatoes can only be attained with those that were 93% mature (McGlasson, 1978 and 1985).

In shelf life improvement, maintaining low ethylene concentration and reducing ethylene biosynthesis plays a significant role. Adams and Yang (1977) have shown that aminoethoxyvinylglycine (AVG) block the conversion of sadenosyl- methionine to 1-aminocyclopropane-1-carboxylic acid. Aminoethoxyvinylglycine was also effective in inhibiting ethylene synthesis in slices of green tomatoes, but was only relatively effective in pink and red tomato fruits. Low temperature also reduces induction of ethylene in tomato fruits during storage, so that the shelf life of tomatoes is increased, when stored at low temperature. Higher temperatures increase ethylene production and result in advanced physiological and biochemical changes in fruit (Wiley, 1994). Ethylene-induced formation of isocoumarin was also characterised in relation to ethylene-enhanced respiration in whole or cut carrots (Lafuente et al., 1996). Sarker and Phan (1979) reported that ethylene induces the formation of isocoumarin (8-hydroxy-3-methyl-6-methoxy-3,4- dihydro-isocomarin) in carrots, a compound associated with bitterness in carrots (Carlton et al., 1961; Simon, 1985). Concentrations of ethylene ranging from 0.1 to 5 ppm, and temperatures from 1 to 15°C, increased respiration, resulting in a more rapid formation of isocoumarin (Lafuente et al., 1996). It was also shown that exposure to low levels of ethylene (0.5 ppm) for 14 days at 1 or 5°C resulted in isocoumarin contents of 20 and 40 mg/100 g peel, respectively. These levels were sufficient to bring a detectable bitter flavour in intact carrot roots. These results clearly indicated that carrot quality is highly sensitive to ethylene during storage and transportation, suggesting that ethylene concentration as well as its biosynthesis should be controlled during commercial storage of carrots.

The presence of the commonly identified phytohormones and various growth regulators are believed to have an inductive effect on ethylene production of fruits and vegetables during ripening (Abdel-Rahman et al., 1975; Davey and Van Staden, 1978; Ryall and Lipton, 1979; Ryall and Pentzer, 1982). Some chemicals applied to bring about abscission of fruits and vegetables that are important in fruit thinning and mechanical harvesting have been shown to induce ethylene production. These can cause premature ripening in fruits and bitterness in carrots.

## Transpirational Loss

Storage temperature and relative humidity play an important role in the physiological changes of fresh produce including physiological weight loss. Water loss is rapid at low relative humidity, since the vapour pressure difference between the commodity and surrounding air is a driving force for moisture transfer from the wet product to the air. In most of the cases moisture content of fresh fruits and vegetables are very high (usually greater than 70%). Therefore, the air inside the flesh is nearly saturated that is, close to 100% relative humidity. Berg and Lentz (1966) noted that the lower humidity ratio causes desiccation and marked softening of carrots, together with some increase in decay. High relative humidity is therefore desirable for reducing physiological weight loss during storage of fruits and vegetables.

Temperature is the other major environmental factor that considerably affects the postharvest physiological weight loss of stored vegetables (Salunkhe et al., 1991). The commodity temperature is highly dependent on the surrounding air temperature. Usually weight loss from perishable commodities is high if surrounding air temperature, flesh moisture content and temperature are high. Vapour pressure increases as air, flesh temperature and moisture content increases. Depending on the magnitude of temperature gradients and relative humidity of the surrounding air the physiological moisture loss varies. In summary, the most important ways to reduce physiological weight loss are by increasing relative humidity and decreasing storage temperature.

## Postharvest Physiological Disorder

Postharvest physiological disorders affect mainly fruits. Susceptibility to disorders was shown to be dependent on a number of factors, such as maturity at harvest, cultural practices, climate during the growth season, produce size, harvesting, and handling practices. Adverse environmental conditions or a nutritional deficiency during growth and development of fruits and vegetables cause postharvest physiological disorders (Brown, 1973; Eckert et al., 1975; Eckert, 1978a, b and c) They may be classified as low temperature disorders, postharvest physiological disorders and mineral deficiency disorders. Low temperature storage is beneficial because it reduces respiration and metabolic activities. Tropical and subtropical fruit and vegetables require specific ranges of storage temperature. On the other hand, low temperature does not Workneh and Osthoff 9313 reduce all aspects of metabolism to the same extent as it reduces respiration. This could lead to a metabolic imbalance due to an accumulation of reaction by-products and possibly a shortage of substrates. Chilling injury is a disorder long observed in plant tissues, especially those of tropical or subtropical origin. This injury is due to the exposure of plant tissue to temperatures below their critical temperature, which is usually below 15°C (Ryall and Lipton, 1979; Bramlage, 1982; Couey, 1982; Wills et al., 1989). The physical symptom of chilling injury and the lowest safe storage temperature for some fruits and vegetables varies. Pitting of the skin due to the collapse of cell beneath the surface and browning of flesh tissues are some of the common symptoms of chilling injury. Therefore, selection of a proper storage temperature range for fruits and vegetables is a critically important factor in order to maintain the best quality and increase shelf life. The approximate lowest safe storage temperature for tomato fruit varies between 7.2-12.8°C (Hardenburg et al., 1986), while carrot roots are not susceptible to chilling injury when stored at temperatures as low as 0°C.

## Chemical and Biochemical Changes During Ripening and Storage

During ripening of fruits, several biodegradation processes take place, such as depolymerization, substrate utilization, loss of chloroplasts, and pigment distraction, mainly due to the action of hydrolytic enzymes (esterases, dehydrogenases, oxidases, phosphatases and ribonucleases) (Baker, 1975; Mattoo et al., 1975). There are also some biosyntheses associated with these processes such as syntheses of proteins and nucleic acids, maintenance of mitochondria, oxidative phosphorylation, phosphate ester formation and syntheses linked to the metabolic pathway (Baker, 1975; Mattoo et al., 1975).

## Carbohydrates

Among the changes that may occur during ripening of flesh fruits, such as tomatoes, is a change in the carbohydrates composition mainly due to substrate utilization and action of hydrolytic enzymes (Pratt, 1975). The presence of free or combined sugars with other constituents plays an important role in flavours of vegetables through a sugar to acid ratio balance. During ripening of fruit, carbohydrates undergo metabolic transformations, both qualitatively and quantitatively. Starch is completely hydrolysed into glucose, and fructose and sucrose formed as ripening progresses (Mattoo et al., 1975). However, structural carbohydrates are decreased slightly. Mattoo et al. (1975) reported that pentosans and cellulose are stored carbohydrates, which may also serve as potential sources of acids, sugars, and other respiratory substances during ripening.

In tomatoes natural sugars such as arabinose, rhamnose, and galactose are steadily decreased in cell walls as the color changes (Campbell et al., 1990). Several studies have also shown that changes in cell wall carbohydrates similar to those reported for intact fruits occurred in cell walls from pericarp discs (model for intact tomato fruits) (Gross and Wallner, 1979; Gross, 1984; Campbell et al., 1990). Other studies showed that the total sugar content of tomatoes increased during ripening, and may be followed by no further changes or a slight decrease during ripening (Baldwin et al., 1991). Similar findings were reported by other researchers, who showed that sucrose content in tomatoes also decreased with the progress of ripening (Goodenough et al., 1982; Baldwin et al., 1991).

During storage of fruit and vegetables, free sugars show a general initial increase followed by a decrease. The increase in free sugars in some fruit and vegetables are due to the breakdown of polysaccharides. In some fruits approximately equal quantities of glucose and fructose are formed due to hydrolysis of starch. However, as storage time advances, especially in fruit, the content of all three free sugars (sucrose, glucose and fructose) declines. Several factors contribute towards the excessive decline of sugars during storage, such as fruit maturity, storage temperature, concentration of $O_2$, $N_2$, ethylene and $CO_2$. Higher temperature favours faster utilisation of sugars as substrate in the respiration process (Wiley, 1994).

In carrots, the ratio of nonreducing to reducing sugars exhibits a sharp decrease after 14 to 18 weeks of storage at 1.1°C Simultaneously an active synthesis of reducing oligosaccharides as raffinose together with the formation of new rootlets is observed (Phan et al., 1973). Temperature also regulates the rate at which biochemical changes occur during storage. Higher temperature activetes enzymatic catalysis and leads to chemical and biochemical breakdown of chemicals in fruits and vegetables during storage. As a result, fruits and vegetables loose firmness faster at higher temperature due to high enzymatic activities (Yoshida et al., 1984)

## Organic Acids

Common acids found in fruit includes citric, malic and ascorbic acid. During ripening, organic acids are among the major cellular constituents undergoing changes (Salunkhe et al., 1991). Studies have shown that there is a considerable decrease in organic acid during ripening of fruits. Modi and Reddy (1966), in Salunkhe et al. (1991), showed that concentrations of citric, malic and ascorbic acids declined 10, 40 and 2.5 fold, respectively, in fruits such as tomatoes. The titratable acidity decreases with storage time, especially at higher temperatures (Mohammed et al., 1999). Higher storage temperatures are known to have an increasing effect on the rate of decrease in ascorbic acid content in tomatoes during storage (Salunkhe et al., 1991). However, Mohammed et al. (1999) showed that the ascorbic acid content in tomatoes slightly increased during ripening during storage at 20°C for 14 and 21 days. In general, the ascorbic acid content decreases rapidly after full ripening of tomatoes stored at higher storage temperatures.

In carrots, titratable organic acidity decreases slowly during storage (Phan et al., 1973). However, Phan et al. (1973) reported that stored carrots had high contents of iso-citric and malic acids, which they could not explain. Pyruvic and oxaloacetic acids were not detected in these stored carrots.

## Enzyme Activity

As was mentioned above, the biochemical changes are responsible for development of off-flavours, discoloration and loss of firmness of fruit and vegetables and are affected by enzymes. The most important enzymes related to food quality include lipolytic acyl hydrolase, lipoxygenase, peroxidase, catalase, protease, polyphenol oxidase, amylase, pectin methylesterase, polygalacturonase, ascorbic acid oxidase and thiaminase (Svensson, 1977). Lipolytic acyl hydrolase, lipoxygenase, peroxidase, catalase and protease are responsible for the changes in flavour of minimally processed as well as intact fruits and vegetables, while polyphenol oxidase is responsible for the changes in colour, especially during ripening of fruits and vegetables (Wiley, 1994). The changes in softness of fruits and vegetables are due to the activities of amylase, pectin methylesterase and polygalacturonase, which result in loss of texture. The reaction catalysed by ascorbic acid oxidase and thiaminase leads to the loss of nutritional quality of food in terms of vitamins C and B content (Wiley, 1994). Catalase and peroxidase are also known for their oxidative activity during ripening and their activity levels increases considerably during ripening of fruit and vegetables (Mattoo et al., 1975).

Glycolytic enzymes are groups of enzymes responsible for the glycolytic breakdown of sugars, and they increase considerably during ripening of fruits. Other classes of enzymes active during storage of fruit and vegetables include hydrolytic enzymes, invertase, transminases, citrate cleavage enzymes, enzymes of the tricarboxylic acid cycle and chlorophyllase (Mattoo et al., 1975), which are responsible for the hydrolysis of starch, flesh softening in most fruits, the degradation of sucrose, liberation of monosaccharides, turnover of amino acids, and ensuing chlorophyll degradation during ripening.

El-Zoghbi (1994) reported effects of enzyme activity on alcohol-insoluble solid dietary fibres, and texture and firmness changes during ripening of tropical fruit such as mango, guava, date and strawberry. The results showed that alcohol-insoluble solid, dietary fibres, texture and firmness declined in these fruit during ripening. However,both polygalacturonase as well as cellulase activities of the fruits increased markedly, during ripening.

The pectin-degrading polygalacturonase (PG) isoenzymes are highly responsible for softening of fruit (Hobson, 1964; Tigchelaar et al., 1978; Marangoni et al., 1995). Bruinsma et al. (1990) demonstrated that these enzymes are highly responsible for most chemical and biochemical changes associated with deterioration of fruit and vegetable quality during ripening and storage. Postharvest treatment had also an effect on biochemical changes related to enzymatic activities during storage (Marangoni et al., 1995; Yoshida et al., 1984; Dodds et al., 1996; Assi et al., 1997). PG normally accumulates to high levels in ripe tomatoes (Kramer et al., 1992). Similarly, Brummell and Harpster (2001) reported that polygalacturonase activity is largely responsible for pectin depolymerization and solublization of polysaccharides. The loss of firmness of tomatoes during ripening is due to the increased activities of PG and pectinesterase (El- Zoghbi, 1994).

In tomato it was shown that ethylene production was induced prior to polygalacturonase production and that it is responsible for triggering PG synthesis indirectly (Grierson and Tucker, 1983). Extracted PG activity in non-chilled tomato fruit significantly correlated with softening of the fruit. However, chilling-associated softening correlated with higher initial extracted pectinmethylesterase (PME) activity. It was suggested that loss of turgor from translocation of water to the PME-modified cell walls was responsible for loss in firmness as a consequence of chilling. Heat treatment, on the other hand, strongly reduces PG activity of tomatoes during ripening and storage (Yoshida et al., 1984). In general, relatively higher temperatures (about 22°C) favour development of PG activity in tomatoes and high temperature treatment suppresses this enzyme activity (Yoshida et al., 1984).

Biochemical processes are also responsible for colouration and loss of firmness of fruits and vegetables (Kertesz, 1951; Doesburg, 1965; Hildebrand, 1989). As a result of increased enzymatic activity occurring due to storage temperature, postharvest quality of tomatoes such as red colour formation, firmness, titratable acidity and decay during ripening are affected (Batu, 1998). Changes in colour from green to red are delayed in the case of tomatoes stored at lower temperature as compared with relatively higher temperature, since the lower storage temperature reduces enzymatic activity. This could lead to increased acidity during storage.

## Postharvest Microbiology of Fruit and Vegetables

The freshness quality and consumer safety of fruits and vegetables depend on the microbial population at harvest, as well as during storage (Brackett, 1992). Fruit and Workneh and Osthoff 9315 vegetables are highly susceptible to microbial contamination during growth, harvest and postharvest operations (Madden, 1992). Most fruits have pH values below 4.5. Consequently, under such acidic conditions most of the microorganisms cannot grow easily. Conversely, most vegetables have pH's above 4.5, which creates favourable conditions for various types of microorganisms. Carrots have pH values varying from 4.9-6.3 while tomatoes have lower pH values ranging between 3.9 and 4.7 (Banwart, 1989). Microorganisms account for up to 15% of postharvest decay of fruits and vegetables. Bacteria, yeast and moulds are the three important groups of microorganisms that affect the quality of fruits and vegetables during storage. The sources of these microorganisms could be from environmental air, soil, and poor sanitation during postharvest unit operations.

Several distinct mechanisms for preharvest infection are known. Lesions on stems, leaves and flower parts of infected plants as well as dead plant materials are sites for sporulation of pathogenic fungi (Nelson, 1965; Salunkhe et al., 1991). Rain and wind transport spores of these fungi to flowers and fruits at every stage of their development. When free water is present and the climatic conditions favourable, these spores germinate and rapidly increase the microbial population. The population and type of microorganisms associated with fruit vegetables are different from those of stem and root vegetables, the reason being the differences in chemical composition such as moisture content, titratable organic acid, and free sugar content. Usually the microbial population is higher on stem and root vegetables than in most fruits mainly due to acidity and may be due to the contact with soil.

The most important postharvest diseases of tomatoes include alternaria, rhizopus, buckeye, grey mold, soft rot, acid rot, bacterial soft and ripe rot, which are caused by Alternaria alternata, Phytophthora species, Botrytis cinerea, Rhizopus stolonifer, Alternaria tenuis, Geotrichum candidum, Erwinia cartovora or Pseudomonas species and Colletotrichum species, respectively (Senter et al., 1987; Splittstoesser, 1987; Buick and Damoglou, 1987; Brackett 1988a; Bulgarelli and Brackett, 1991; Salunkhe et al., 1991). The major postharvest diseases of fresh carrots are gray moulds rot, centrospora rot, watery soft rot and bacterial soft rot which is due to Botrytis cinerea, Centtrospora acerina, Sclerotinia sclerotiorium and E. cartovora, respectively. Similarly, losses of other vegetables in general are due to watery soft rot, cottony leak, fusarium rot, and bacterial soft rot that are caused by *Sclerotinia, Pythium butleri, Fusarium and Erwinia or Pseudomonas* species, respectively.

As ripening of fruits and vegetables progresses, the firmness decreases and the intrinsic factors, which confer resistance during development, can no longer protect against microbial decay (Eckert et al., 1975, 1978). The high moisture and nutrient content of fruits and vegetables adds to susceptibility to invasion by specific pathogenic microorganisms (Eckert, 1978 a, b and c).

The microbial deterioration of fruits and vegetables are highly influenced by environmental factors, such as temperature, moisture content of crops, relative humidity of air, and storage air gas composition. In general, control of postharvest decay of fruits and vegetables are based upon the prevention of infection before as well as after harvest, eradication of incipient infections and retarding the progress of the pathogen in the host by postharvest treatments such as disinfection (Eckert et al., 1975).

# POST-HARVEST HANDLING AND STORAGE

## Pre-Packaging Treatments

### *Postharvest Disease Control*

As mentioned above, the postharvest quality and safety of fresh fruits and vegetables mainly depends on their microbial population (Brackett, 1992). Different approaches have been used to reduce postharvest loss, classified in chemical and non-chemical methods. Various synthetic fungicides and bactericides have been used alone or in combination with chlorination to limit the growth and transmission of diseases caused by microorganisms during storage of fresh commodities. However, chemicals may have a negative impact on agriculture in developing countries (Eckert, 1990; Conway et al., 1999; Qadir and Hashinaga, 2001). Biological products such as Aspire, BioSave, and Yield Plus are now available in the marketplace of developed countries to control growth of bacteria or fungi. *Pseudomonas syringae van Hall* (strain ESC-11) is sold commercially under the names BioSave- 11 and BioSave-110 (EcoScience Corp., Orlando, Florida), and is currently being used to control decay on apples and pears caused by *B. cinerea, Mucorpiriformis fischer*, and *Pseudomonas expansum* (Janisiewicz and Marchi, 1992; Janisiewicz and Jeffers, 1997).

Other methods used to control postharvest diseases are modification of pH, chemical treatments or a combination (Huxsoll and Bolin, 1989; King and Bolin, 1989). The success of these products, however, remains limited (Wisniewski et al., 2001). The reason may include the following: (a) there is an increasing consumer concern over the use of pesticide as a postharvest decay control of food (Wisniewski and Wilson, 1992), (b) the use of fungicides may result in predominance of fungicide resistant strains (Elmer and Gaunt, 1994), (c) the variability experienced in the efficiency of these products, and lack of understanding of how to adapt "biological approaches" to a commercial setting (Wisniewski et al., 2001). Researchers and industries have come up with some noble ideas such as using combinations of different antagonists. However, these approaches seem to meet with consumer resistance (Wisniewski et al., 2001).

Much work as been done on non-chemical approaches and efficient postharvest disease control methods including various physical treatments such as biocontrol agents, heat, ultra-violet light and ionising radiation. Some novel heat treatment approaches to postharvest decay control of fruits and vegetables were developed and applied in Israel (Rodov et al., 1995; Afeke et al., 1998; Fallik et al. 1999, Fallik et al., 2000). The postharvest decay of bell pepper was significantly decreased by hot water treatment dipping at temperatures of 45 and 53°C for 15 and 4 min, respectively (Gonzalez et al., 1999).

Heat treatment, when combined with fungicide treatments, was also reported to be effective and can be an useful alternative method to control postharvest decay (Conway et al., 1999). In this case the effect of fungicide residues on human health and costs associated with combinations of treatments has to be considered. However, the application of some of these methods, such as heat treatment, require equipment that is complex and costly for use in under-developed countries (Rodov et al., 1996).

Intensive research was conducted on the efficiency of chlorine solutions to control the growth of pathogenic microorganisms on fresh produce and minimally processed "ready-to-eat" fruits and vegetables (Wei et al., 1995; Zhuang et al., 1995; Park et al., 1995; Velazques et al., 1998; Beuchat et al., 1998; Delaguis et al., 1999; Escudere et al., 1999; Li et al., 2001; Beuchat et al., 2001; Ukuku and Sapers, 2001; Prusky et al., 2001). These studies were carried out on the survival of inoculated spoilage fungi such as *Alternaria hydrophila* (Velazquez et al., 1998; Beuchat et al., 2001), *A. alternata* (Prusky et al., 2001) and human pathogenic bacteria such as *Salmonella montevideo* (Zhuang et al., 1995), *Salmonella stanley* (Ukuku and Sapers, 2001) and *Yersinia enteroocolitica* (Escudero et al., 1999). Similarly, studies on the normal microbial populations of harvested vegetables were also reported (Beuchat et al., 1998; Prusky et al., 2001). The concentration of free chlorine in these studies varied from 50 to 2000 and ppm and dipping times from 1 to 10 min.

Chlorination, when used in combination with nonchemical methods, is effective in reducing postharvest disease losses. Warm chlorinated water was shown to reduce microbial loads by 3 log cfu.

$g^{-1}$ in lettuce washed in chlorinated water at 47°C, and 1 log cfu. $g^{-1}$ at 4°C (Delaquis et al., 1999). Similarly, the combination of different chlorine concentrations (50 to 400 mg/ml NaOCl, pH 7.0) with 52.5°C water were shown to be more effective than use of water or chlorine solutions at 20°C for reducing initial microbiological flora in minimally processed green onions Park et al. (1995) reported that the microbial load of the minimally processed watercress and onion was effectively reduced with chlorine (100 ppm). However, high concentrations of chlorine resulted in greater microbial proliferation after 7 days of storage, loss of ascorbic acid and significant colour changes in stored cut vegetables.

Preharvest dipping treatment in the organic chlorine compound troclosene sodium extended the storage life of fruit by delaying development of black-spot disease (Prusky et al., 2001). All the research work cited here, has been devoted to the control of pathogens and postharvest decay by microorganisms without any emphasis on the effect of free chlorine on the quality, or the physiological and biochemical quality changes.

In general, chlorination is an effective method especially in reducing microbial load in intact fruits and vegetables. For several reasons, such as availability of chlorinated solutions and cost, chlorine disinfection seems to be a good method for application in developing countries. The efficacy of chlorine solution on postharvest physiological as well as chemical and biochemical changes still needs to be explored.

## Electrochemically Activated water (Anolyte Water)

Anolyte water is prepared from an aqueous solution of NaCl and this electrochemically-activated water is known to be a powerful, non-toxic, non-hazardous disinfectant. Two kinds of electrochemically activated water are produced, anolyte and catholyte. The anolyte water is described as having an oxidation-reduction potential (ORP) in the region of +1000 mV and catholyte an ORP of –800 mV, and the pH value of catholyte is in the alkaline region while the pH value of anolyte is in the acidic region. Current thinking around the concept is that the ORP of both solutions fluctuates between these values at a rate too rapid to measure. Some of the biocidal agents (free radicals) in the solutions are $ClO_2$, HClO, $Cl_2$, ClO-, $H_2O_2$, $HO_2$ -, NaHO, $O_2$, $O_3$, H· and ·OH. Prilutski and Bakhir (1997) investigated properties of anolyte and catholyte produced by diaphragm electrolysis of aqueous solutions. Their results showed that catholyte and anolyte have different physical-chemical properties. Anolyte is thought to have the antimicrobial effect and catholyte a detergent or cleaning effect (Popova et al., 1999). The presence of the free radicals with their oxidising effects in the solutions is considered of great importance. It is the free radicals (working substances), which gives anolyte bactericidal and sporicidal activity (Aquastel, 2000), since higher organisms possess antioxidant defence systems, whereas microorganisms generally do not.

The electrochemically produced anolyte water was effectively used in drinking water disinfecting units (Aquastel, 2000). The determination of organic chlorine compounds and trihalogenmethanes showed significantly less formation of these by-products. Moreover, the formation of halogens were significantly less than defined standard requirements and the pH-value only slightly decreased (Aquastel, 2000). Anolyte attracts interest since it is an environmentally and ecologically friendly substance for use as a postharvest fruit and vegetables disinfectant. Based on these and the antimicrobial action Workneh and Osthoff 9317 of anolyte, its potential to be used as a postharvest disinfectant of fruit and vegetables should be explored.

## Packaging

### *Controlled Atmosphere Packaging*

Controlled atmosphere storage refers to storage of food in a gas atmosphere that is different from the normal composition of air. In controlled atmosphere the headspace gas is precisely adjusted to the atmosphere composition required by a specific fruit or vegetable (Perry, 1993). Controlled atmosphere storage is mostly used for the long-term storage of whole fruits and vegetables in warehouses, bulk storage and transportation.

Over 4000 research papers have been published giving information on the optimal modified atmosphere conditions for cultivars of fruits and vegetables (Zagory and Kader, 1988; Kader et al., 1989). These literatures have built good basic principles of controlled atmosphere storage and significantly contributed towards the success of this technology in developing countries (Lioutas, 1988). The major influence on the choice of controlled atomsphere conditions is the susceptibility of particular cultivars to superficial scald (Johnson, 1999).

Using controlled atmosphere storage methods, the gas compositions have to be adjusted to a certain level specific to each fruit and vegetable type, and it is unlikely that there will be benefits from further adjustment of the compositions of oxygen and carbon dioxide without risking the damage to certain fruits and vegetables (Johnson, 1999).

The controlled atmosphere gas compositions required by a given commodity is obtained by mixing appropriate volumes of gases. Certain machinery are needed to mix and control the gas composition of the air during the storage period. In controlled-atmosphere storage packaging, atmospheric conditions (level of gases) are continuously controlled and adjusted to maintain the optimal concentrations required by specific fresh produce. As a result of this, it is capital intensive and expensive to operate for short-term storage (Kader et al., 1989). The factors that could probably determine the success of controlled atmosphere technology in developing countries are the cost and complexity (multidisciplinary nature) of the technology. However, controlled atmosphere storage is not appropriate for some commodities. For example, Lipton (1977) has shown that controlled atmosphere storage was not advantageous for carrot storage, and recommended the use of modified atmosphere packaging.

Low $O_2$ or high $CO_2$ levels may be of benefit for very short-term preservation of vegetables, as for example, with asparagus (Torres-Penarada and Saltveit, 1994), banana (Wills et al., 1990) and some vegetables (Wills et al., 1979). Low $O_2$ or high $CO_2$ conditions result in partial or full anaerobic respiration and produce volatile compounds (acetaldehyde and ethanol) responsible for aroma of fresh fruits (Morris et al., 1979; Paz et al., 1981). However, excessive accumulation of volatile compounds may lead to off-flavours and some physiological disorders (Ke and Kader, 1989). A study on the effects of low $O_2$ and high $CO_2$ levels on tomato physiology and composition showed that fruits treated with low $O_2$ or high $CO_2$ had skin injury and blotchy ripening (Klieber et al., 1996). The incidence of fruits showing disease when ripe, increase with increasing treatment time with $O_2$ and $CO_2$

(Klieber et al., 1996). Sozzi et al. (1999), on the other hand, reported that keeping tomatoes in controlled atomspheres, even in the presence of ethylene, had marked residual effects. These results suggested an antagonism between elevated $CO_2$ or low $O_2$, and exogenous ethylene levels, which could determine most of the physiological behaviour under controlled atmosphere storage, though a direct regulatory mechanism by $O_2$ and/or $CO_2$ should not be discarded. The literature cited indicated that controlled atmosphere storage of tomatoes in conditions of low $O_2$ and $CO_2$ atmosphere might have economic benefits for very short-term storage of a few days.

## Modified Atmosphere Packaging (MAP)

Historically, modified atmosphere packaging (MAP) has been erroneously described as synonymous with controlled- atmosphere storage (CAS) or controlled-atmosphere packaging (CAP). Modified atmosphere packaging (MAP) is defined as the "packaging of perishable products in an atmosphere which has been modified so that its composition is other than that of air" (Hintlian and Hotchkiss, 1986). MAP storage implies a lower degree of control of gas concentrations as compared with controlled atmosphere and vacuum packaging. Typically, initial atmospheric conditions are established for a transient period, and the interplay of the commodity physiology and the physical environment maintain those conditions within broad limits (Zagory and Kader, 1988). Modified atomsphere is depletion of oxygen ($O_2$) and emanation $CO_2$, occurring within a sealed plastic packaging headspace.

This happens when fruit and vegetables are respiring. Modified atmosphere packaging can be created either as a result of commodity respiration (passive) or intentionally through active packaging (Zagory and Kader, 1988). Passive modified atmosphere can be achieved in hermetically sealed packages due to commodity respiration. Levels of $O_2$ and $CO_2$ concentrations then adjusted themselves to the commodity-desired range as a result of $O_2$ consumption and $CO_2$ evolution during storage. If a commodity's respiration characteristics are matched to the film permeability value, the optimum favourable modified atmosphere can be created for a specific product (Smith et al., 1987).

The atmosphere created by modified atmosphere packaging (MAP) depends on product, environmental and packaging film factors. Product factors include respiration rate, respiration quotient, quantity of the product, and $O_2$ and $CO_2$ concentrations necessary to approximately achieve optimum reduction of product aerobic respiration rate. Packaging film factors consists of permeability of polymeric packaging materials to $O_2$ and $CO_2$, and water vapour at a selected storage temperature. Temperature, relative humidity, light and sanitation represent the environmental factors. This indicates that as many factors as possible should be included during MAP of fresh vegetable trials (Zagory and Kader, 1988). A proper understanding of the basic principles of the physiology, chemical, biochemical and microbiological properties of vegetables during MAP storage is required for the use of this technology.

Physiological changes: The beneficial effect of modified atmosphere packaging on increasing shelf life of perishable products is partially due to the decrease in $O_2$ and the increase in $CO_2$ levels, and partially to the decrease in physiological loss in weight (Biale, 1946, 1960; Fidler et al., 1973; Isenberg, 1979; Smock, 1979; Ben-Yehoshua et al., 1983). The role of MAP is to primarily reduce respiration rate of fruit and vegetables by retarding metabolic activities. Reduced respiration also retards softening, and slows down various compositional changes associated with ripening (Kader, 1986).

It is obvious that plant tissue responds quickly to low $O_2$ or high $CO_2$ levels during MAP storage. Low $O_2$ levels were shown to reduce the rate of respiration and resulted in a delay in climacteric onset of the rise in ethylene levels and a decrease in the rate of ripening (Fidler et al., 1973; Kader, 1986; Kannelis, et al., 1990; Yang and Chinnan, 1988a, b; Kannelis et al., 1991). Hypoxia conditions (low $O_2$ levels) have shown to inhibit the accumulation of RNA, protein, and DNA synthesis associated with the wounding of vegetables (Butler et al., 1990). The lowest limit of $O_2$ level, which can be tolerated by vegetables, is a critical factor that must be taken into consideration in selecting proper MAP for fruits and vegetables such as carrots and tomatoes (Butler et al., 1990).

Burg and Burg (1967) have shown that the $CO_2$ level is a competitive inhibitor of ethylene production during storage, and metabolic activities are reduced through an increase in $CO_2$ levels in the headspace. Low $O_2$, high $CO_2$ or both decreases the production of ethylene during modified atmosphere packaging (MAP) storage, which means lower rates of respiration, and prolongs the shelf life of produce (Blackman, 1954; Aligue, 1995; Agar and Streif, 1996; Petracek et al., 2002). This is for the benefit of both climacteric and non-climacteric fruits and vegetables during storage. The levels of $CO_2$ and $O_2$ in the headspace of packages balance each other and have a coupled effect on the metabolic activities of fresh produce during storage. The effect of lower $O_2$ and higher $CO_2$ levels on respiration is additive and can reduce respiration more than effected separately (Kader et al., 1989).

These means high levels of $CO_2$, when coupled with low $O_2$ levels have a cumulative effect on the respiration rate.

Oxygen and $CO_2$ levels below or above the optimum levels specific to each fruit or vegetable can cause physiological damage. Exposure of fresh commodities to extreme levels of $O_2$ and $CO_2$ below and above that required for normal respiration, may initiate anaerobic respiration and lead to the development of off-flavours due to the accumulation of ethanol and acetaldehyde (Zagory and Kader, 1988). Increased $CO_2$ is also associated with an increase of acidity in plant tissues with a subsequent reduction in pH. This acidification could lead to inactivation of enzymes, or phytotoxicity by $CO_2$.

Chemical and biochemical changes: The loss in firmness, development of off-flavours, and discoloration of fresh and minimally processed commodities after harvest or during ripening or storage, is the result of various biochemical changes. A review on the biochemical basis for effects of modified atmosphere packaging (MAP) on fruits and vegetables by Kader (1986) indicated that biochemical and compositional changes such as color, texture (firmness), flavour, and nutritive value can be reduced due to the modified atmosphere created by MAP during storage. MAP may inhibit enzymatic activities responsible for the deterioration in the mentioned quality parameters of fresh commodities during storage. The loss of chlorophyll and lycopene synthesis is reduce when tomatoes are stored in a modified atmosphere of less than 4% $O_2$ and 4% $CO_2$ for 2 months (Goodenough and Thomas, 1980). The appearance of PG is delayed for up to 8 weeks in green mature tomatoes stored in 5% $O_2$ and 5% $CO_2$ at 12.5°C (Goodenough et al., 1982), suggesting delayed fruit softening. Burton (1974) reported that the changes in starch to sugar conversion is slower in packaged fruit and vegetables with a headspace containing 5 to 20% $CO_2$ or less than 3% $O_2$ at 2°C.

Under MAP the effect of polyphenoloxidase and tyrosinase, the enzymes responsible for browning of plant tissue, decreases as a result of limited availability of $O_2$ as a substrate (Anese et al., 1997). A reduction in $O_2$ levels or increase in $CO_2$ levels in the microenvironment of a crop by modified atmosphere packaging (MAP) could reduce enzymatic activities responsible for quality deterioration (Duckworth, 1966; Ruall and Lipton, 1979; Ryall and pentzer, 1982; Kader et al., 1985). However, the effect of low $O_2$ created during MAP on biochemical changes cannot be separated from the effect of increased $CO_2$ (Murr and Morris, 1974; Bown, 1985; Siriphanich and Kader, 1986; Ballantyne et al., 1988b). Barth et al. (1993) reported that MAP of broccoli stored at 20°C for 96 h resulted in a significant decrease of peroxidise (POX) activity after 24 h compared to POX activities in unpackaged samples. The chemical compositions such as ascorbic acid and total chlorophyll also remained higher in packaged broccoli, but $O_2$ levels below 3% increases the Workneh and Osthoff 9319 concentrations of ethanol and acetaldehyde, and activities of pyruvate decarboxylase and alcohol dehydrogenase, which results in ethanolic fermentation in fresh-cut carrots (Kato-Noguchi and Watada, 1997).

This review shows that MAP has an effect on the biochemical as well as chemical changes in fruits and vegetables during ripening and storage through reducing the levels of $O_2$ and increasing the levels of $CO_2$ in the headspace. Generally, the lower the $O_2$ concentrations achieved by the MAP the lower the rate of chemical as well as biochemical changes, resulting in reduced respiratory metabolism and other biochemical processes, and leading to retention of chemical quality characteristics.

However, maintenance of optimum minimum $O_2$ concentration inside the packages is very important to avoid accumulation of excessive volatile compounds responsible for off-flavour and physiological disorder. This entails the need for selection of appropriate packaging material for the specific fruits and vegetables.

Microbiological changes: The extended shelf life of many MAP may allow extra time for microorganisms to reach dangerously high levels in a food. Even in the case of slow growing photogenic microorganisms, extended shelf life using MAP gives them sufficient time to multiply, if not properly managed during storage. Additional research in the microbial control on MAP foods is still needed.

Hao et al. (1999) studied microbiological quality and production of botulinum toxin in MAP of broccoli, carrots, and green beans. Types of packaging material affected the quality of vegetables without noticeably influencing the growth of microorganisms. These results created awareness that the use of modified atmosphere packaging (MAP) could have both positive as well as negative effects on the microbial flora associated with fruits and vegetables during storage based on the pre-packaging treatment and storage conditions. However, their result showed that the toxin was not detected in the vegetables under study.

Microorganisms require certain conditions to grow and reproduce. These conditions are either properties of food such as nutrients, pH and water activity or extrinsic properties of food associated with storage environment such as oxygen, light, temperature and time. The most important environmental factors are gaseous composition and temperature. Modified atmosphere packaging, in principle, controls these factors to reduce or suppress spoilage and extend shelf life of fresh produce. Anese et al. (1997), studying the effect of modified atmosphere packaging on microbial growth, have shown that $CO_2$ and $N_2$ levels were the most effective in inhibiting microbial growth. Concentrations of $CO_2$ in excess of 5% inhibit the growth of microorganisms of most foods. However, the pH of MAP foods increases with increase in carbon dioxide concentration in packages. Direct suppression of microorganisms by MAP has only been possible with the horticultural commodities that can withstand carbon dioxide levels at 20% or above without development of off-flavours, odours, and colours (Harvey, 1982; Aaron, 1989). The ability of modified atmosphere alone is not sufficient to limit the incidence of spoilage of fruits and vegetables. The necessity of pre-packaging treatments to reduce initial microbial load, as well as limiting their growth during storage, is critical.

Conversely, unsuitable MAP can prevent wound healing, speed up senescence, or incidence of physiological breakdown, making fruits and vegetables more susceptible to postharvest pathogens. EL-Goorani and Sommer (1981) suggested that $O_2$ levels below 1% of $CO_2$ levels above 10% could be utilised to limit the growth of yeasts and moulds.

In summary, the growth of microorganisms is likely to continue inside MAP vegetables and fruits, especially for those that need higher $O_2$ and lower $CO_2$ levels. Additional pre-packaging treatments are therefore still required.

## Postharvest Factors Affecting MAP

The microenvironment created and maintained in the headspace of a package is the net result of the interaction of commodity factors, environmental factors and packaging material characteristics. Zagory and Kader (1988) described the effects and development of modified atmospheres in the packaging of fresh produce. In order to maintain quality of fruits and vegetables, the product, environmental and packaging material factor must be considered in creating favourable conditions for MAP of fresh produce in order to keep the quality. Therefore, in this section a brief description of commodity factors and environmental and packaging material factors for MAP will be presented. Commodity factors: In order to develop an appropriate modified atmosphere environment in packages several product factors need to be understood. These includes the rate of $O_2$ uptake and $CO_2$ evolution, permeability of packaging films to $O_2$ and $CO_2$, tolerance of plant materials to levels of these gases and diffusivity of fruits and vegetable skin and flesh to $O_2$ and $CO_2$ (Fidler et al., 1973; Isenberg, 1979; Smock, 1979; Kader, 1985; Salveit, 1989). Zagory and Kader (1988) classified the commodity factors affecting the effectiveness of MAP according to resistance to gas diffusion, respiration, ethylene production and sensitivity, optimum temperature, relative humidity, and gas concentrations.

Resistance to gas and water diffusion across the plant material varies with different vegetables or fruits, cultivars, plant organs and stage of maturity (Cameron and Reid, 1982). The resistance to gas diffusion is affected more by anatomical difference than biochemical differences among various vegetables and fruits (Burton, 1974).

However, the rate at which $O_2$ passes through skin or epidermal tissue depends on the $O_2$ gradient (Kader et al., 1989; Burg, 1990). In some commodities, the skin and epidermal tissue may possess pores, called stomata, and lenticles, which are the passages for the $O_2$. Cranberry, tomato and green pepper skins contain no pores, and all gases diffuse through holes in the pedicel scar. The resistance of gas diffusion through the boundary gas layer and skin of most fruits and vegetables varies with the type of gases, that is, $O_2$, $CO_2$ and ethylene (Burg, 1990). The rate of gas diffusion from the surface of the product to the interior also depends on the proportion of gas space. In the case of minimally processed fruits and vegetables, the released cellular fluid flows into the gas space and blocks gas transfer. Diffusion of gas would be much slower in cell sap than in the gas-filled intercellular spaces (Burg, 1990). In general, the structures such as the skin, the cell wall, the intercellular space and the cytosol are the barriers of $O_2$ movement to the mitochondria in the cell.

The respiration of fruits and vegetables was discussed above, but without taking into account packaging films and MAP. The effectiveness of packaging films can be evaluated by their ability in keeping the normal respiration rate of produce without decomposition. Based on the rate of respiration, fruits and vegetables are classified as climacteric or non-climacteric. Tomato and carrot are classified as climacteric fruit and root vegetables, respectively, with moderate respiration rates compared to the other fruits and vegetables. The respiration rate increases consistently with an increase in storage temperature. The respiration rate of topped carrots is from 10 to 20 mg $kg^{-1}$ $h^{-1}$ and 46 to 95 mg $kg^{-1}$ $h^{-1}$ at 0°C and 20 to 21°C respectively. For mature green tomatoes the respiration rate varies from 5-8 and 28-41 mg $kg^{-1}$ $h^{-1}$ at 4 to 5°C and 20 to 21°C storage (Salunkhe et al., 1991). Thus, it is apparent that a reduction in storage temperature is an effective means of extending the commercial life of tomatoes and carrots using MAP.

The rate of ethylene production of fresh produce, and the sensitivity to this hormone, determines the type of packaging film appropriate for secured quality maintenance and shelf life extension (Zagory and Kader, 1988). Tomato produces 1 to 10 mg $CO_2$. $kg^{-1}$ $h^{-1}$ ethylene, which is moderate compared to some other fruits (Kader et al., 1985). Very low concentrations of ethylene are required to advance biochemical and physiological changes during storage and accelerate the ripening processes of carrots and tomatoes. The effect of ethylene on fruits and vegetables that can be stored at a temperature range of 0 to 4.4°C is not possible to detect. The recommended storage temperatures for carrots range from 0 to 1°C and hence the effect of ethylene on carrot during storage at this temperature is negligible. The recommended storage temperature for tomatoes is varying from 8 to 13°C, which is sufficiently high for production of ethylene, and should be considered as a postharvest factor affecting the shelf life of tomatoes.

A reduction in ethylene production and sensitivity associated with MAP can delay the onset of climacteric and prolong the storage life of these fruits (Zagory and Kader, 1988). Even non-climactreric fruits and vegetables can benefit from reduced ethylene sensitivity and lower respiration rate attributed to MAP. Ethylene production is reduced by ether low $O_2$, high $CO_2$, or both, and the effects are additive (Zagory and Kader, 1988).

Environmental factors: Air temperature, relative humidity and natural microbial flora of perishable produce are the main environmental factors affecting the effectiveness of modified atmosphere packaging (Zagory and Kader, 1988). Packaged fruits and vegetables have specific ranges of temperature optima for their normal respiration. Below this limit, fruits and vegetables can easily by affected by chilling injury, which advances to high respiration, hastened senescence and short shelf life. The optimum storage temperature for a specific fruit or vegetable is the one that delays senescence, and maintains quality the longest without causing chilling, freezing, or other injury (Zagory and Kader, 1988). The optimum temperature for tropical fruits is above 13°C, because these fruits are sensitive to chilling injury. Nonchilling- sensitive commodities can be stored near 0°C without negative effect. Previous studies indicated that the effect of reduced $O_2$ and elevated $CO_2$ had an effect on chilling injury (Lyons and Breidenbach, 1987). Chilling injury of fruits and vegetables reduced with low $O_2$ and high $CO_2$ concentrations in the storage atmosphere (Lyons and Breidenbach, 1987). The optimum temperature range for carrot and tomatoes are 0-5 and 8-13°C, respectively (Salunkhe et al., 1991). In general, optimum temperature is a key factor in reducing respiration, production of ethylene, chilling injury, enzymatic activities, proliferation of microorganisms during storage of fruits and vegetables, although modified atmosphere packaging expands the limit of optimum temperature.

The danger associated with high relative humidity is the condensation in packages, which creates favourable conditions for microbial growth. Condensation on the package film affects the permeability property and creates unfavourable interior conditions during storage. Proper temperature maintenance throughout the postharvest handling lines is central to prevent the incidence of condensation in films. On the other hand, low relative humidity increases the loss of moisture from fruits and vegetables (Kader, 1987). The optimum range of relative humidity for carrot storage varies from 90 to 100%, whereas for tomatoes, the range can vary from 85 to 90% (Salunkhe et al., 1991)

## Packaging Films

Selection of appropriate films for MAP storage of fresh vegetables on their effectiveness in protecting produce Workneh and Osthoff 9321 during shelf life is critical (Arthey and Dennis, 1991; Perry, 1993; Wiley, 1994; Farber and Dodds, 1995). The principal plastic materials for MAP that can be used with intact vegetables and fruits include polybutylene, low density polyethylene (LDPE), high density polyethylene (HDPE), polypropylene (PP), polyvinylchloride (PVC), polystyrene (PS), ethylene vinyl acetate (VA), ionomer, rubber hydrochloride (pliofilm), and polyvinylidine chloride (PVDC) (Schlimme and Rooney, 1994). Permeability of packaging film to $O_2$ and $CO_2$ varies with type of packaging film material or composition. The characteristics of the main types of plastic films with their potential uses in MAP are summarized in Table 1 (Schlimme and Rooney, 1994). The table shows that the transmission rate for $O_2$ varies from 8 up to as high as 13,000 $cm^3\ m^{-2}$ $day^{-1}$ at 1 atm for PVC and LDPE packaging films, respectively. The transmission rate of packaging materials for $CO_2$ also varies from 59 up to 77,000 ($cm^3\ m^{-2}\ day^{-1}$ at 1 atm), which is quite a large variation, the lowest permeability being for PVDC and highest for LDPE. The water vapour transmission rate also varies from 1.5 to 155 g m-2 $day^{-1}$. The variation in permeability of packaging films to $O_2$, $CO_2$ and $H_2O$ could give a wide range of choices for films appropriate for specific commodities.

Temperatures have effects on the permeability of packaging films. As temperature increases, the film permeability increases, which may affect the passive modified atmosphere that the film is designed for. Permeability of $CO_2$ is affected more than $O_2$ permeability, implying that a film that is a suitable packaging material at one temperature may not be so at another (Zagory and Kader, 1988). Kader et al. (1989) reported that most common films are relatively good barriers to moisture vapour, because they maintain high internal humidity even in dry, ambient conditions. This means that relative humidity affects the permeability of packaging films less than storage temperature. However, condensation inside films decreases the permeability of $CO_2$ and $O_2$ (Rodov et al., 1995).

It is obvious that the quality of fresh commodities can be dependent on the types of packaging film used for MAP. Under poor modified atmosphere storage management, occurrence of anaerobic metabolism is also possible in commodities packaged in films with low permeability to $O_2$, $CO_2$ and water vapour. Losses in physiological weight are higher from fresh commodities packaged in low-density packaging films than in highdensity films. However, because near ambient $O_2$ level is maintained in the headspace of packages, the occurrence of anaerobic metabolism of fresh commodities could be avoided with the use of packaging films with higher permeability to $O_2$, $CO_2$ and water vapour during storage. Lower weight loss and firmness of tomatoes were maintained better in polyolefin bags (0.015 mm thick) than in PVC (0.0177 mm thick) during 5 weeks of storage at 10°C and 80% relative humidity (Frezza et al., 1997).

**Table 1:** Permeability characteristics of several plastic films with potential for use as MAP of fresh and lightly processed produce (Schlimme and Rooney, 1994).

| **Film type** | **Transmission rate** | | |
|---|---|---|---|
| | **$O_2$*** | **$CO_2$*** | **$H_2O$ vapour**** |
| Low density polyethylene (LDPE) | 3900 to 13000 | 7700 to 77000 | 6 to 23.2 |
| Linear low density polyethylene (LL DPE) | 7000 to 9300 | - | 16 to 31 |
| Medium density polyethylene (MDPE) | 2600 to 8293 | 7700 to 38750 | 8 to 15 |
| High density polyethylene (HDPE) | 52 to 4000 | 3900 to 10000 | 4 to 10 |
| Polypropylene (PP) | 1300 to 6400 | 7700 to 21000 | 4 to 10.8 |
| Polyvinylchloride (PVC) | 620 to 2248 | 4263 to 8.138 | >8 |

| | | | |
|---|---|---|---|
| Polyvinylchloride (PVC), plasticized | 77 to 7750 | 770 to 55000 | >8 |
| Polystyrene (PS) | 2000 to 7700 | 10000 to 26000 | 108.5 to 155 |
| Ethylene vinyl acetate copolymer (12% VA) | 8000 to 13000 | 35000 to 53000 | 60 |
| Ionomer | 3500 to 7500 | 9700 to 17800 | 22 to 30 |
| Rubber hydrochloride (Pliofilm) | 130 to 1300 | 520 to 5200 | >8 |
| Polyvinylidine chloride (PVDC) | 8 to 26 | 59 | 1.5 to 5 |

*Expressed in terms of $cm^3\ m^{-2}\ day^{-1}$ at 1 atm. **Expressed in terms of $g\ m^{-2}\ day^{-1}$ at 37.8°C and 90% RH.

The shelf life of tomatoes was extended better when packaged in polyethylene and PP films as compared with PVC bags (Batu and Thompson, 1998). The lowest weight losses and the highest soluble solids contents after 60 days of storage were found in tomatoes packaged in polyethylene and PP films. Shelf life studies of pink ripe tomatoes packaged in polyethylene films with 100, 200 or 300 gauge showed that freshness was best maintained in perforated films compared with the other (Singh et al., 1996). The reason for this may be due to higher permeability to gases of the two films as compared with polyvinyl chloride.

MAP using LDPE delayed changes in acidity, soluble solids, texture, colour and PG activity, and resulted in substantial reductions of weight loss of tomatoes, while higher density films showed the lowest weight losses, and the highest soluble solids content after 60 days of storage (Nakhasi et al., 1991). However, it could be possible that the higher microbial populations found in these packages were due to relatively a higher storage temperature (13°C) and moisture accumulation in the headspace of packages.

Ben-Yehoshua et al. (1998) has shown that perforating the film affected the concentrations of both $O_2$ and $CO_2$ and water condensation without affecting the relative humidity of the air in the package headspace. The perforated film prevented the decay and maintained the postharvest quality of vegetables and fruits compared to storage in open boxes. Polyethylene bags with 2 to 4 diffusion holes per 4 kg tomatoes gave the best results, as the $CO_2$ permeability was improved (Esquerra and Bautista, 1990). Similarly, LDPE bags were found to be better in maintaining the quality of carrots during storage compared with high-density polypropylene bags (Seyoum and sadik. 2001). This indicated that the microperforated packaging films maintain the best quality of tomatoes during MAP. The strategies for combining films or film and micropore combinations could lead to realistic solutions to meet the required gas transfer properties (Exama et al., 1993).

The common films, such as PVC and LDPE, can satisfy the gas flux requirements of low respiring tissues of carrots (Exama et al., 1993). Emond et al. (1995), on the other hand, showed that perforated films significantly reduced water loss and condensation in packages of carrots during storage at 2°C and 95% relative humidity, compared to unperforated films. Nazar et al. (1996) showed that perforations in polyethylene bags equivalent to 0.5% of the total effective surface area of the package had no effect on the shelf life of carrots. The packaging film permeability also greatly influences disease incidence and sprouting of carrots during storage (Lingaiah and Reddy, 1996), as did non-ventilated polyethylene bags (Lingaiah and Reddy, 1996).

## SUMMARY

The literature survey shows that any preharvest treatment, either to improve yield or quality, has an effect on fresh commodity quality at harvest and during storage. It is clear that not all the reviewed conditions could be optimised simultaneously. The reviewed conditions were also not described for specific cultivars of a vegetable type. This means that many of these conditions will have to be accepted as constant in a study, e.g., agricultural practices and environmental conditions, with focus on a few selected variables, e.g. ComCat® treatment, postharvest disinfection and storage temperature. The postharvest performance of high yield ComCat® treated vegetables, e.g. carrots and tomatoes have not yet been investigated. The reviewed literature on postharvest performance of carrots and tomatoes stretches over several years and the research was carried out in many countries, while using different experimental approaches. Experiments will have to be designed carefully and proper controls will have to be incorporated.

Postharvest treatments of vegetables include disinfecting, and use of MAP and cold storage. According to the literature survey, disinfecting methods can gene--rally be classified as chemical, biological, and physical methods. Although, several chemicals and biological products are available in the market, their success remains limited due to variability experienced in their effectiveness and lack of experience on how to adapt them to the commercial settings. The literature shows that chlorine treatment is currently the most commonly used, easiest to use and the cheapest method, but not necessarily the most effective. It is very effective for control of microorganisms, but almost no knowledge is available on its effect on the physiological aspects of harvested vegetables. Chemical methods have residues that have negative impacts on health and environment or ecology, while physical methods need high initial investment. This encourages further research not only for microbiological safety, but also to develop environmentally and ecologically safe sustainable disinfectants as a partial or complete substitution for chemicals.

Regarding extended shelf life, the literature pointed to MAP and specific storage temperature for best results. New packaging material is constantly reaching the market, so that those aspects should be the final methods to test the postharvest performance of preharvest treated vegetables. MAP, combined with cold storage are the most popular method as a result of its efficiency in increasing the shelf life of fresh commodities without much quality deterioration, less operational and material cost and easy consumer packaging. The observation that preharvest treatment increase harvest yield of vegetables may be effectively used for food security purposes in underdeveloped countries when production is immediately followed with sustainable preservation technology. Despite high market demand after production, 20 to 50% of commodities are lost before reaching consumers in these countries due to a shortage of cooling facilities.

# REFERENCES

1. Abdel-Rahman M, Thomas TH, Doss GJ, Howell L (1975). Changes in endogenous plant hormones in cherry tomato fruits during development and maturation. Physiol. Plant, 34: p. 39.

2. Adams DO, Yang SF (1977). Methionine metabolism in apple tissue. Plant Physiol. 60: p. 892.

3. Afek U, Orenstein J, Nuriel E (1998). Increased quality and prolonged storage of sweet potatoes in Israel. Phytoparasitica, 26: 307-312.

4. Agar IT, Streif J (1996). Effect of high $CO_2$ and controlled atmosphere (CA) storage on the fruit quality of raspberry. Gartenbauwissenschaft, 66(6): 261-267.

5. Aligue R (1995). Residual effects of short term treatments with high $CO_2$ on the ripening of cherimoya (*Annona cherimola* Mill.) fruit. J. Workneh and Osthoff 9323 Hort. Sci. 70(4): 609-615.

6. Anese M, Manzano M, Nicoli MC (1997). Quality of minimally processed apple slices using different modified atmosphere conditions. J. Food Qual. 20: 359-370.

7. Aquastel (2000). Superior sterilisation, disinfecting and water purification. Eurostel water disinfecting systems. Institute für Hygiene und Umwelt, IHU, Steinstrasse 10, D-33457 Lollar, Germany.

8. Arthey D, Dennis C (1991). Vegetable Processing. London, Blackie.

9. Assi NE, Huber DJ and Brecht JK (1997). Irradiation-induced changes in tomatoes fruit and pericarp firmness, electrolyte efflux, and cell wall enzyme activity as influenced by ripening stage. J. Am. Soc. Hort. Sci. 122(1): 100-106.

10. Baker JE (1975). Morphological changes during maturation and senscence, in Postharvest physiology, Handling and Utilization of Tropical and Subtropical Fruits and Vegetables, Pantastico EB, Ed., AVI Publishing, westport, CT.

11. Baldwin EA, Nisperos MO, Moshonas MG (1991). Quantitative analysis of flavour and other volatiles and for certain constituents of two tomato cultivars during ripening. J. Am. Soc. Hort. Sci. 116(2): 265-269.

12. Ballantyne A, Stark R, Selman JD (1988b). Modified atmosphere packaging of broccoli florets. Int. J. Food Sci. Technol. 23: 353-360. Banwart GJ (1989). Basic Food Microbiology. Van Nostrand Reinhold, 115 Fifth Avenue, New York, 10003.

13. Barth MM, Kerbel EL, Broussard S, Schmidt SJ (1993). Modified atmosphere packaging protects market quality in broccoli spears under ambient temperature storage. J. Food Sci. 58(5): 1070-1072.

14. Batu A (1998). Effect of storage temperature and modified atmosphere packaging on ripening quality and chilling injury of mature green tomatoes. Bahce. 26(1-2): 67-77.

15. Batu A, Thomposon AK (1998). Effects of modified atmosphere packaging on post harvest qualities of pink tomatoes. Turk. J. Agric. For. 22(4): 365-372.

16. Ben-Yehoshua S, Shapiro B, Even-Chen Z, Lurie S (1983). Mode of action of plastic film in extending life of lemon and bell pepper fruits by alleviation of water stress. Plant Physiol. 73: 87-93.

17. Berg LVD, Lentz CP (1966). Effect of temperature, relative humidity, and atmospheric composition on changes in quality of carrots during storage. Food Technol. 20(7): 104-107.

18. Beuchat LR, Nail BV, Adler BB, Clavero RS (1998). Efficiency of spray application of chlorinated water in killing pathogenic bacteria on row apples, tomatoes and lettuce. J. Food Prot. 61(10): 1305-1311.

19. Beuchat lR, Harris LJ, Ward TE, Kaja TM (2001). Development of a proposed standard method for assessing the efficacy of fresh produces sanitizers. J. Food Prot. 64(8): 1103-1109.

20. Biale JB (1946). Effect of oxygen concentration on respiration of avocado fruit. Am. J. Bot. 33: 363-373.

21. Biale JB (1960). Respiration of fruits. In Handbuch Der Plantephysiologie. Encyclopedia of Plant Physiology, Vol. XII/2, J. Wolf (ed), pp. 536-592. Berlin: springer-Verlag.

22. Bialczyk J, Lechowski Z, Libik A (1996). Fruiting of tomato cultivated on medium enriched with bicarbonate. J. Plant Nutr. 19: 305-321.

23. Blackman FF (1954). Analytical studies in Plant Respiration. London: Cambridge University Press.

24. Bown AW (1985). $CO_2$ and intracellular pH. Plant Cell Environ. 8: 459- 465.

25. Brackett RE (1988a). Changes in the microflora of packaged fresh tomatoes. J. Food Qual. 11: 89-105.

26. Brackett RE (1992). Shelf stability and safety of fresh produce as influenced by sanitation and disinfection. J. Food Prot. 55(10): 808- 814.

27. Bruinsma J, Geelen TAM, Knegt E, Varga A, Vermeer E (1990). Postharvest development of climacteric fruit. Proceedings of the international congress of Plant Physiology, New Delhi, India, 15-20 February, 1988. Vol. 2: 1385-1391.

28. Brummell DA, Harpster MH (2001). Cell wall metabolism in fruit softening and quality and is manipulation in transgenic plants. Plant Mol. Biol. 47(1-2): 311-340.

29. Buick RK, Damoglou AP (1987). The effect of vacuum packaging on the microbial spoilage and shelf life of 'ready-to-use' sliced carrots. J. Sci. Food Agric. 38: 167-175.

30. Bulgarelli MA, Brackett RE (1991). The importance of fungi in vegetables. Food and Feeds. Vol. 3 p. 179. Marcel Dekker, New York.

31. Burg SP, Burg EA (1967). Molecular requirements for the biological activity of ethylene. Plant Physiol. 42: 144-152.

32. Burg SP (1990). The theory and practice of hypobaric storage, pp.353- 372. In Food Preservation by Modified Atmospheres. Calderon M and Barkai-Golan R (eds.). CRC Press, Boca Raton, FL. Burton WG (1974). Some biophysical principles underlaying the controlled atmosphere storage of plant material. Ann. Appl. Biol. 78: 149-168.

33. Butler WC, Cook C, Vaya ME (1990). Hypoxic stress inhibits multiple aspects of potato tuber wound process. Plant Physiol. 93: 264-270.

34. Cameron AC, Reid MS (1982). Diffusive resistance: Importanance and measurment in controlled atmosphere storage, pp. 171-181. In: Richardson DG, Meheriuk M (eds). Proc. Third Natl. Controlled Atmospherees Res. Conf. Timber Press, Portland. Ore.

35. Campbell AD, Hugsamer M, Statz HU, Greve LC, Labavitch JM (1990).

36. Comparison of ripening processes in intact tomato fruit and excised pericarp disc. Plant Physiol. 94: 1582-1589.

37. Carlton BC, Peterson CE, Tolbert NE (1961). Effects of ethylene and oxygen on production of a bitter compound by carrot roots. Plant Physiol. 36: 550-552.

38. Carmer MD, Oberhozer JA, Combrink NJ (2001). The effect of supplementation of root zone dissolved inorganic carbon on fruit yield and quality of tomatoes (cv 'Daniella') grown with salinity. Scientia Hortic. 89: 269-289.

39. Clouse SD (1996). Molecular genetic studies confirm the role of brassinosteroids in plant growth and development. Plant J. 10: 1-8. Colla G, Casa R, Lo Cascio B, Saccardo F, Leoni C (1999). Fertilization in cenetral Italy. Acta Hort. 487: 531-535.

40. Conway WS, Janisiewicz WJ, Klein JD, Sams CE (1999). Strategies for combining heat treatment, calcium infiltration, and biological control to reduce postharvest decay of Gala apples. Hort. Sci. 34(4): 700- 704.

41. Couey HM (1982). Chilling injury of crops of tropical and subtropical origin. Hort. Sci. 17(2): 162-165.

42. Dalal KL, Salunkhe DK, Boe AA, Olson LE (1965). Certain physiological and biochemical changes in the developing tomato fruit. J. Food Sci. 30: 504-508.

43. Davey JE, Van Staden J (1978). Endogenous cytokinins the fruits of ripening and non-ripening tomatoes. Plant Sci. Lett. 11: p. 359. Davies PJ (1995). Plant Hormones: Physiology, Biochemistry and Molecular Biology. Dordrecht: Kluwer.

44. Delaquis PJ, Stewart S, Toivonen PMA, Moyls AL (1999). Effect of warm, chlorinated water on the microbial flora of shredded iceberg lettuce. Food Res. Int. 32: 7-14.

45. Dodds GT, Trenholm L, Madramootoo CA (1996). Effect of watertable and firtilizer management on susceptibility of Tomato fruit to chilling injury. J. Am. Soc. Hort. Sci. 121(3): 525-530.

46. Doesburg JJ (1965). Pectic substances in fresh and preserved fruits and vegetables. I.V.V.T. Communication NR. 25. The Netherlands: Institute for Research on Storage and Processing of Horticultural Produce Wageningen.

47. Duckworth RB (1966). Fruits and Vegetables, Pergamon Press, Oxford. England.

48. Eckert JW (1990). Recent developments in the chemical control of postahrvest diseases. Acta Hort. 269: 477-494.
49. Eckert JW (1978a). Pathological deseases of fresh fruits and vegetables. In Postharvest Biology and Biotechnology, Hultin HO and Milner M Eds. Food and Nutrition Press, Westport, C.T. p. 161
50. Eckert JW (1978b). Pathological disease of fresh fruits and vegetables. J. Food Biochem. 2: p. 243.
51. Eckert JW (1978c). Postharvest diseases of fresh fruits and vegetables-etiology and control, in Postharvest Biology and Handling of Fruits and Vegetables, Haard NF and Salunkhe DK Eds. AVI Publishing Westport, CT.
52. Eckert JW, Rubia PP, Mottoo AK, Thompson AK (1975). Diseases of tropical crops and their control in Postharvest Physiology, Handling, and Utilization of Tropical and Subtropical Fruits and Vegetables. Pantastico EB Ed. AVI Publishing, Westport, C.T. 415.
53. El-Goorani MA, Sommer NF (1981). Effect of modified atmospheres on pathogens of fruits and vegetables. In: Jenick J (ed.). Hort. Rev. The AVI publishing company Inc., Westport, Conn. 3: 412-461.
54. Elmer PAG, Gaunt RE (1994). The biological characteristics of dicarbiximide-resistant isolates of *Manilinia fructicola* from New Zealand stone acid. Plant Physiol. 98: 995-1002.
55. Emond JP, Boily S, Mercier F (1995). Reduction of water loss and condensation using perforated film packaging for fresh fruits and vegtables. Am. Soc. Agric. Eng. pp. 339-346.
56. Escudero ME, Velazquez L, DiGenaro MS, de Cortinez YM, de Guzman AM (1999). Elimination of Yersinia enterocolitica by chlorine on fresh tomatoes. Central European J. Public Health, 7(1): 24-26.
57. Esquerra EB, Bautista OK (1990). Modified atmosphere storage and transport of 'Improved Pope' tomatoes. ASEAN Food J. 5(1): 27-33.
58. Evers AM (1989a). Effects of different fertilization practices on the glucose, fructose, sucrose, taste, taste and texture of carrot. J. Agric. Sci. Finl. 61: 113-122.

59. Exama A, Arul J, Lencki R, Li Z (1993). Suitability of various plastic films for modified atmosphere packaging of fruits and vegetables: gas transfer properties and effect of temperature fluctuation. Acta Hort. 343: 175-180.

60. Fallik E, Aharoni Y, Copel A, Rodov V, Tuvia-Alkalai S, Horev B, Yekutieli O, Wiseblum A, Regev R (2000). A short hot water rinse reduces postharvest losses of Galia melon. Plant Pathol. 49: 333-338.

61. Fallik E, Grinberg S, Alkalai S, Yekutieli O, Wiseblum A, Regev R, Beres H, Bar-Lev E (1999). A unique and fast postharvest method to improve storage quality of sweet pepper. Postharvest Biol. Technol. 15: 25-32.

62. Farber JM, Dodds KL (1995). Principles of Modified and Sous Vide Product Packaging Lancaster, P.A: Technomic Publishing Co., pp 464.

63. Fidler JC, Wilkinson BG, Edney KL, Sharples RO (1973). The biology of apple and pear storage. Res. Rev. 3, Commonwealth Agric. Bur. England, p. 235.

64. Frenkel C (1972). Involvement of peroxidase and indole-3-acetic acid oxidase isozymes from pear, tomato, and blue berry fruit in ripening. Plant Physiol. 49: p. 757.

65. Frezza D, Moccia S, Trinchero G, Fraschina A, Chiesa A, Yehoshua S (1997). MAP: polyolefin and pVC films on the quality and shelf life of tomatoes. 14th International congress on plastics in agriculture, Tel Aviv, Israel, March 1998. pp. 513-520.

66. Gao ZM, Sagi M, Lips H (1996). Assimilate allocation priority as affected by nitrogen compaounds in the xylem sap of tomato. Plant Physiol. Biochem. 34: 807-815.

67. Gonzalez-Angilar GA, Cruz R, Baez R (1999). Storage quality of bell peppers pre-treated with hot water and polyethylene packaging. J. Food Qual. 22: 287-299.

68. Goodenough PW, Thomas TH (1980). Comparative physiology of fieldgrown tomatoes during ripening on the plant or retarded ripening in controlled atmospheres. Ann. Appl. Biol. 98: 507.

69. Goodenough PW, Tucker GA, Grierson S, Thomas T (1982). Changes in colour, polygalacturonase monosacharides and organic acids during storage of tomatoes. Phytol. Chem. 21(2): 281-284.

70. Grierson D, Tucker GA (1983). Timing of ethylene and polygalacturonase synthesis in relation to the control of tomato fruit ripening. Planta, 157: 174-179.
71. Gross KC (1984). Fractionation and partial characterisation of cell walls from normal and non-ripening mutant tomato fruit. Physiol. Plant, 62: 25-32.
72. Gross KC, Wallner SJ (1979). Degradation of cell wall polysaccharides during tomato fruit ripening. Plant Physiol. 63: 117-120.
73. Hao X, Papadopoulos AP (1999). Effects of supplemental lighting and cover materials on growth, photosynthesis, biomass partitioning, early yield and quality of greenhouse cucmber. Scientia Hort. 80: 1- 18. Hao YY, Brackett RE, Beuchat LR, Doyle MP (1999). Microbiological quality and production of botulinal toxin in film-packaged broccoli, carrots, and green beans. J. Food Prot. 62(5): 499-508.
74. Hardenburg RE, Warada AE, Wang CV (1986). The Commercial Storage of Fruits, Vegetables, and Florist and Nursery Stocks, Agriculture Handbook No 66, USDA, Washington. D.C.
75. Harvey John M (1982). $CO_2$ Atmospheres for Truck Shipments of Strawberries. Controlled Atmospheres for Storage and Transport of Perishable Agricultural Commodities. Ore. State Univ. School of Agr. Sym. Ser. No. 1: 359-365.
76. Hildebrand DF (1989). Lipoxygenases. Physiol. Plant, 76: 249-253.
77. Hintlian CB, Hotchkiss JH (1986). The safety of modified-atmosphere packaging. J. Food Technol. 40: 70-76.
78. Hobson GE (1964). Polygalacturonase in normal and abnormal tomato fruit. Biochem. J. 92: 324-332.
79. Hobson GE, Davies JN (1971). The Tomato. In: Hulme AC (ed.). The Biochemstry of fruits and their Products, Vol. 2, Academic Press, NY, pp. 437-482.
80. Hole CC, McKee JMT (1988). Changes in soluble carbohydrate levels and associated enzymes of field-grown carrots. J. Hort. Sci. 63(1): 87-93.
81. Hüster T (2001). Personal communication. AgraFurUm, Germany.

82. Huxsoll CC, Bolin HR (1989). Storage stability of minimally processed fruit. J. Food Process. Preserv. 13(4): 281-292. Isenberg MFR (1979). Controlled atmosphere storage of vegetables. Hort. Rev. 1: 337-394.

83. Janisiewicz WJ, Jeffers SN (1997). Efficacy of commercial formulation of two biofungicides for control of blue mold and gray mold of apples in cold storage. Crop Prot. 16: 629-633. Janisiewicz WJ, Marchi A (1992). Control of storage rots on various pea cultivars with a saprophyte strain of *pseudomonas syringae*. Plant Dis. 76: 555-560.

84. Johnson DS (1999). Controlled atmosphere storage of apple in UK. Proceedings of the International Symposium on Effect of Preharvest and Postharvest Factors on Storage of fruits. Acta Hort. 2(485): 187-193.

85. Kader AA (1985). Postharvest biology and technology: An overview, pp. 3-7. In Postharvest Technology of Horticultural Crops. Kader AA,

86. Kasmire RF, Mitchell FG, Reid MS, Sommer WF and Thompson JF (eds.). Special Publication. 3311, University of California, Davis, CA.

87. Kader AA (1986). Biochemical and physiological basis for effects of controlled and mdified atmospheres on fruits and vegtables. Food Technol. 40(5): 99-104.

88. Kader AA (1987). Postharvest biology and technology research on deciduous tree fruits: recent trends and future outlook. In: Proc. Summerland Research Station. Commemorative Symposium (April 27-28, 1987). Agric. Canada, Summerland, B.C. Canada. pp. 96- 105.

89. Kader AA, Kashmire RF, Mitchell FG, Reid MS, Sommer NF, Thompson JF (1985). Postharvest Technology of Horticultural Crops. University of California Division of Agriculture and Natural Resource, Davis, CA.

90. Kader AA, Zagory D, Kerbel EL (1989). Modified atmosphere packaging of fruits and vegetables. CRC Rev. Food Sci. Nutr. 28(1): 1-30.

91. Kannelis AK, Solomos T, Roubelakis-Angelakis KA (1991). Suppression of cellulase and polygalacturonase and induction

of alcohol dehydrogenase isoenzymes of avocado fruit mesocarp subjected to low oxygen stress. Plant Physiol. 96: 269-274.

92. Kannelis AK, Solomos T, Roubelakis A (1990). Suppression of cellulase and polygalacturonase and induction of alcohol dehydrarognase isoenzymes in avocado fruit mesocarp subjected to lo oxygen stress. Plant Physiol. 96: 269-277.

93. Kato-Noguchi H, Watada AE (1997). Effect of low-oxygen atmosphere on ethanolic fermentation in fresh-cut carrots. J. Am. Soc. Hort. Sci. 122(1): 107-111.

94. Ke D, Kader AA (1989). Tolerance and response of fresh fruits to low oxygen levels at or below 1%. Proceedings Fifth International Conference Controlled Atmospheres, Vol. 2 Wenatchee, Washington, pp. 209-216.

95. Kertesz ZI (1951). The pectic substances. New York-London: Interscience.

96. King Jr. AD, Bolin HR (1989). Physiological and microbiological storage stability of minimally processed fruits and vegetables. Food Technol. 43(2): 132-135.

97. Klieber A, Ratanachinakom B, Simons DH (1996). Effects of low Workneh and Osthoff 9325 oxygen and high carbon dioxide on tomato cultivar 'Bermuda' fruit physiology and composition. Scientia Horticulturae, 65(4): 251-261.

98. Knee M, Finger FL (1992). NADP+-malic enzyme and organic acid levels in developing tomato fruits. J. Am. Soc. Hort. Sci. 117(5): 799- 801.

99. Kramer M, Sanders R, Bolkan H, Waters C, Sheehy RE, Hiatt WR (1992). Postharvest evaluation of transgenic tomatoes with reduced levels of polygalacturonase: processing, firmness and disease resistance. Postharvest Biol. Technoil. 1(3): 241-255.

100. Lafuente MT, Lopez-Galvez G, Cantwell M, Yang SF (1996). Factors influencing ethylene-induced isocoumarin formation and increased respiration in carrots. J. Am. Soc. Hort. Sci. 121(3): 537-542.

101. Leonardi C, Guichard S, Bertin N (2000). High vapour pressure deficit influences growth, transpiration and quality of tomato fruits. Scientia Hort. 84: 285-296.

102. Li Y, Brackett RE, Shewfelt RL, Beuchat LR (2001). Changes in appearance and natural microflora on iceberg lettuce treated in warm, chlorinated water and then stored at refrigeration temperature. Food Microbiol. 18: 299-308.

103. Lieblein G (1993). Quality and yield of carrots: effect of composet manure and mineral fertilizer. Agricultural University of Norway, Doctor Scietarum thesis no. 13.

104. Lipton WJ (1977). Recommendations for CA storage of broccoli, a brussels sprouts, cabbage, cauliflower, asparagus and potatoes. Horticultural Report No. 28, July 1977. Proceedings of the Second National Controlled Atmosphere Research Conference, April 5-7,Michigan State University.

105. Lopez S, Maroto JV, San Bautista A, pascual B, Alagarda J (2000). Qualitative changes in pepino fruits following preharvest application of ethephon. Scientia Hort. 83(2): 157-164.

106. Lyons JM, Breidenbach RW (1987). Chilling injury. In:Postharvest Physiology of Vegtables, ed Weichman J., p. 305. Marcel Dekker. Inc., New York.

107. Lyons JM, Pratt HK (1964). Effect of stage of maturity and ethylene treatment on respiration and ripening of tomato fruits. Proc. Am. Soc. Hort. Sci. 84: p. 491.

108. Marangoni AG, Jackman RL, Stanley DW (1995). Chilling-associated softening of tomato fruit is related to increased pectinmethylesterase activity. J. Food Sci. 60(6): 1277-1281.

109. Mattoo AK, Murata T, Pantastico EB, Chachin K, Ogata K, Phan CT (1975). Chemical changes during ripening and senscence, in Postharvest physiology, Handling and Utilization of Tropical and Subtropical Fruits and Vegetables, Pantastico EB Ed. AVI Publishing, westport, CT.

110. Mc Glasson WB (1978). Role of hormone in ripening and senscence. In: Hultin HO and Milmer M (eds). Postharvest Biology and Biotechnology, Food and Nutrition Presses, Westport C.T., pp. 77- 96. McGlasson WB (1985). Ethylene and fruit ripening. Hort. Sci. 20: 51-54.

111. Mitchell CA, Chun C, Brandt WE, Nielsen SS (1997). Environmental modification of yield and nutrient composition of Waldmann's Green leaf lettuce. J. Food Qual. 20: 73-80.

112. Mohammed M, Wilson LA, Gomes PI (1999). Postharvest sensery and physiochemical attributes of processing and nonprocessing tomato cultivars. J. Food Qual. 22: 167-182.

113. Morris JR, Cawthon DL, Buescher RW (1979). Effect of acetaldehyde on postharvest quality of mechanically harvested strawberries for processing. J. Am. Soc. Hortic. Sci. 104: 262-264.

114. Mortensen LM (2000). Effect of air humidity on growth, flowering, keeping quality and water relations of four short-day greenhouse species. Scientia Hort. 86: 299-310.

115. Mozafar A (1994). Organic Fertilizers May Increase Some Plant Vitamins. In Transactions. 15th World Congress of Soil Science (Acapulco, Mexico, July 10-16 1994). Chapingo; Mexico: Sociedad Mexicana de la Ciencia del Suelo, pp. 656-666.

116. Murr DP, Morris LL (1974). Influence of $O_2$ and $CO_2$ on o-diphenol oxidase activity in mushroom. J. Am. Soc. Hort. Sci. 99: 155-158.

117. Nelson LW (1965). Harvest practices that increase sweet potato surface rot in storage. Phytol. Pathol. 55: p. 640.

118. Ozeker FS, Karacali I, Yildiz M, Kinay P, Yildiz F (2001). The effect of some preharvest treatments on wound healing in satsuma mandarin after harvest. Proceedings of the Fourth International Conference on

119. Postharvest Science. Acta Hort. 1(553): 73-75.

120. Paez L, Hultin HO (1972). Respiration of potato mitochondria and whole tuber and relationship to sugar accumulation. J. Food Sci. 35: p. 46Pardossi A, Bagnoli G, Malorgio F (1999). NaCl effect on celery

121. (*Apium graveolens* L.) grown in NFT. Scientia Hort. 81: 229-242.

122. Park W, Lee D, Park WP, Lee DS (1995). Effect of chlorine treatment on cut water cress and onion. J. Food Qual. 18(5): 415-424.

123. Paz O, Janes HW, Prevost BA, Frenkel C (1981). Enhancement of fruit sensory quality by postharvest applications of acetaldehyde and ethanol. J. Food. Sci. 47: 270-273, 276.

124. Perry RT (1993). Principles and Applications of modified Atmosphere Packaging of Foods. Blackie Academic and Professional, An Imprint of Chapman and Hall, Wester Cleddens Road, Bishopbriggs, Glasgow G64 2NZ, UK.

125. Petracek PD, Joles DW, Shirazi A, Cameron AC (2002). Modified atmosphere packaging of sweet cherry (*prunus auium* L., ev. Sams) fruit: metabolic responses to oxygen, carbon dioxide, and temperature. Postharvest Biol. Technol. 24(3): 259-270.

126. Petrichenko VN, Romanova AV, Savitskaja AK (1996). Carrot losses during storage depend on the growth conditions. Kartofel'i Ovoshchi. 3: 2-13.

127. Popova I, Yu Kiselev VI, Lobyshev VI (1999). On the nature of biological activity of water and aqueous solutions after treatment in the diaphragm electolyser. Moscow pp31-46.

128. Pratt HK (1975). The role of ethylene facteurs et regulation de la maturation des fruits. Cent. Nut. Rec. Sci. p. 153.

129. Prusky D, Eshel D, Kobiler H, Yakoby N, Moualem DB, Ackerman M, Zuthji Y, Arie RB (2001). Postharvest chlorine treatments for the control of the persimmon black spot disease caused by *Alternaria alternata*. Postahrvest Biol. Technol. 22(3): 271-277.

130. Qadir A, Hashinaga F (2001). Inhibition of postharvest decay of fruits by nitrous oxide. Postharvest Biol. Technol. 289: 279-285.

131. Raven PH, Evert RF, Eichhorn SE (1992). Biology of Plants. New York: Worth. pp. 545-572.

132. Rodov V, Ben-Yehoshua S, Albagi R, Fag DQ (1995). Reducing chilling injury and decay of stored citrus fruits by hot water dips. Postharvest Biol. Technol. 5: 119-127.

133. Rodov V, Peretz J, Agar T, D'hallewin G, Ben-Yehushua S (1996). Heat applications as complete or partial substrate of postahrvest fungicide treatments of grapefruit and oroblanco fruits. Proc. Int. Soc. Citriculture, 2: 1153-1157.

134. Rosen JC, Kader AA (1989). Postharvest physiology and quality maintenance of sliced pear and strawberry fruits. J. Food Sci. 54: 656-659.

135. Rosenfeld HJ (1999). Quality improvement of vegetables by cultural practices. Proceedings of the International symposium on Quality of Fresh and Fermented Vegetables, 1(483): 57-66.

136. Ryall AL, Lipton WJ (1979). Handling, Transportation and Storage of Fruits and Vegetables. Vol. 1 Vegetable and Melons, 2nd ed, AVI Publishing, Westport, CT.

137. Ryall AL, Pentzer WT (1982). Handling, Transportation and Storage of Fruits and Vegetables, Vol. 2. Fruits and Tree Nuts. 2nd ed. AVI Publishing. Westport. CT.

138. Salisbury FB, Ross CW (1992). Plant Physiology. Belmont, CA: Wadsworth. pp. 357-407, 531-548.

139. Salunkhe DK, Bolin HR, Reddy NR (1991). Storage, processing, and nutritional quality of fruits and vegetables. 2nd edition. Volume I. Fresh fruits and Vegetables.

140. Salunkhe DK, Wu M, Wu MT, Singh B (1971). Effect of Telone and Nemagon on Essential nutritive components and the respiratory rates of carrot (*Daucus carota* L.) roots and sweet corn (*Zea mays* L.) seeds. J. Am. Soc. Hort. Sci. 96(3): 357-359.

141. Salveit MS (1989). A summary of requirements and recommendations for the controlled and modified atmosphere storage of harvested vegetables. In: Controlled Atmosphere Research Conference, Wenatchee, WA.

142. Sargent SA, Talbot MT, Brecht JK (1991). Evaluating precooling methods for vegetable packinghouse operations. Veg. Crop Dept.

143. Special Series SSVEC-47, Inst. Food and Agric. Sci. University of Flav, Gainesville, FL.

144. Sarker SK, Phan CT (1979). Naturally-occuring and ethylene-induced phenolic compaounds in the carrot roots. J. Food Prot. 42: 526-534.

145. Sasse J, Smith R, Hudson I (1995). Effect of 24-epibrassinolide on germination of seeds of *Eucalyptus camaldulensis* in saline conditions. 15th International Conference on Plant Growth Substances, Minneapolis, Minnesota, USA.

146. Schlimme DV, Rooney ML (1994). Packaging of minimally processed fruits and vegetables. In Minimally processed refrigerated fruits and vegetables, by Wiley RC (ed.). New York: Chapman and Hall. pp. 135-179.

147. Schnabl H, Roth U, Friebe A (2001). Brassinosteroids-induced stress tolerances of plants. Recent Res. Dev. Phytochem. 5: 169-183.

148. Sen F, Karacali I, Yildiz M, Kinay P, Yildiz F, Iqbal N (2001). The effect of some preharvest treatments on wound healing in satsuma

mandarin after harvest. Proceedings of the Fourth International Conference on Postharvest Science. Acta Hort. 1(553): 77-78.

149. Senter SD, Bailey JS, Cox NA (1987). Aerobic microflora of commerciallly harvestes, transported and cryogenically processed collards (Brassica olearacea). J. Food Sci. 52: 1020-1021.

150. Sasse JM (1997). Recent progress in brassinosteroid research. Physiol. Plant, 100, 696-701.

151. Seyoum TW, sadik K (2000). Natural ventilation evaporative cooling of mango. J. Agric. Biotechnol. Environ. 2(1/2): 1-5.

152. Simon PW (1985). Carrots flavour: effects of genotype, growing conditions, storage and processing. In: pattee HE (ed.). Evaluation of quality of fruits and vegetables. AVI publ. Co., Westport, CT, pp. 315-328.

153. Singh WMB, Salunkhe WDA, Dull GG (1970). Effects of certain soil fumigants on essential nutritive componants and the respiratory rates of carrot (*Daucus carota* L.) roots. Hort. Sci. 5: 221-222.

154. Siriphanich J, Kader AA (1986). Changes in cytoplasmic and vacuolar pH in harvested lettuce tissue as influenced by CO2. J. Am. Soc. Hort. Sci. 111: 73-77.

155. Smith SM, Geeson JD, Stow J (1987). Production of modified atmospheres in deciduous fruits by the use of films and coatings. Hort. Sci. 22: 772.

156. Smock RM (1979). Controlled atmosphere storage of fruits. Hort. Rev. 1: 301-336.

157. Sørensen JN, Jørgensen U, Kühn B (1997). Drought effects on the marketable and nuteritional quality of carrots. J. Sci. Food Agric. 74: 379-391.

158. Sozzi GO, Trinchero GD, Fraschina AA (1999). Controlled atmosphere storage of tomato fruit: low oxygen or elevated carbon dioxide levels alter galactosidase activity and inhibit exogenous ethylene action. J. Sci. Food Agric. 79(8): 1065-1070.

159. Splittstoesser DF (1987). Fruits and fruit products. In Food and beverage Mycology, Beuchat LR (ed.). pp. 101-128, New york: AVI/van Nostrand Reinhold.

160. Svensson S (1977). Inactivation of enzymes during thermal processing. In Physical, Chemical, and Biological Changes in Food Caused by

161. Thermal Processing, Hoyem T and Kvale O (eds.). p. 202 London: Applied Science.

162. Takatsuto S, Kamuro Y, Watanabe T, Noguchi T, Kuriyama H (1996). Proc. Plant Growth Regulator Soc. Am. 23: p. 15.

163. Thomas RL, Jen JJ, Morr CV (1981). Changes in soluble and bound peroxidase-IAA oxidase during tomato fruit development. J. Food Sci. 47: 158-161.

164. Tigchelaar EC, McGlasson WB, Buescher RW (1978). Genetic regulation of tomato fruit ripening. Hort. Sci. 13: 508-513.

165. Tittonell P, De Grazia J, Chiesa A (2001). Effect of nitrogen fertilisation and plant population during growth on lettuce (*lactuca sativa* L.) postharvest quality. Proceedings of the Fourth International Conference on Postharvest Science. Acta Hort. 1(553): 67-68.

166. Torres-Penaranda AV, Saltveit ME (1994). Effect of brief anaerobic exposures on carbon dioxide production and quality of harvested asparagus. J. Am. Soc. Hortic. Sci. 119: 551-555.

167. Ukuku DO, Sapers GM (2001). Effect of sanitizer treatments on Salmonella Stanley attached to the surface of cantaloupe and cell transfer to fresh-cut tissues during cutting practices. J. Food Prot. 64(9): 1286-1291.

168. Velazquez LC, Escudero ME, DiGenaro MS, DeCortinez YM, de Guzman AM (1998). Survival of Aeromonas hydrophila in fresh tomatoes (*lycopersicum esculentum* Mill) stored at different temperatures and treated with chlorine. J. Food Prot. 61(4): 414-418.

169. Watkins CB, Pritts MP (2001). The influence of cultivars on postharvest performance of fruits and vegetables. Proceedings of the Fourth International Conference on Postharvest Science. Acta Hort. 1(553): 59-63.

170. Wei CI, Huang TS, Kim JM, Lin WF, Tamplin ML, Bartz JA (1995). Growth and survival of salmonella montevideo on tomatoes and disinfection with chlorinated water. J. Food Prot. 58(8): 829-836.

171. Wiley RC (1994). Minimally processed refrigerated fruits and vegetables. New York: Chapman and Hall. P1-368.
172. Wills RBH, McGlasson WB, Graham D, Lee TH, Hall EG (1989). Postharvest-An Introduction to the Physiology and Handling of fruit and vegtables, 3rd ed., Van Nostrand Reinhold, New York.
173. Wills RBH, Klieber A, David R, Siridhata M (1990). Effect of brief premarketing holding of bananas in nitrogen on time to ripen. Aust. J. Exp. Agric. 30: 579-581.
174. Wills RBH, Wimalasiri P, Scott KJ (1979). Short pre-storage exposures to high carbon dioxide or low oxygen atmospheres for the storage of some vegetables. Hort. Sci. 14: 528-530.
175. Workneh and Osthoff 9327
176. Wisniewski M, Wilson C, EL-Ghaouth A, Droby S (2001). Non-chemical approaches to postharvest disease control. Acta Hort. (ISHS) 553: 407-412.
177. Wisniewski ME, Wilson CL (1992). Biological control of postharvest diseases of fruits and vegetables: recent advances. Hort. Sci. 27: 94- 98.
178. Yamada H, Kobayashi S (1999). Relationship between watercore and maturity or sorbitol in apples affected by preharvest fruit temperature. Scientia Hort. 80: 189-202.
179. Yang CC, Chinnan MS (1988a). Modelling the effect of $CO_2$ on respiration and quality of stored tomatoes. Am. Soc. Agric. Eng. 31: 920-925.
180. Yoshida O, Nakagawa H, Ogura N, Sato T (1984). Effect of heat treatment on the development of polygalacturonase activity in tomato fruit during ripening. Plant Cell Physiol. 25(3): 505-509.
181. Zagory D, Kader AA (1988). Modified atmosphere packaging of fresh produce. Food Technol. 42(9): 70-77, 76-77.
182. Zhuang RY, Beuchat LR, Angulo FJ (1995). Fate of Salmonella montevideo on and in raw tomatoes as affected by temperature and treatment with chlorine. Appl. Environ. Microbiol. 61(6): 2127-2131.

Chapter 8

# The Importance of Mineral Elements for Humans, Domestic Animals and Plants: A Review

K. O. Soetan[1], C. O. Olaiya [2], and O. E. Oyewole[3]

[1]Department of Veterinary Physiology, Biochemistry and Pharmacology, University of Ibadan, Ibadan, Oyo state, Nigeria

[2]Department of Biochemistry, University of Ibadan, Ibadan, Oyo state, Nigeria

[3]Department of Health Promotion and Education, University of Ibadan, Ibadan, Oyo state, Nigeria

## ABSTRACTS

Minerals are inorganic nutrients, usually required in small amounts from less than 1 to 2500 mg per day, depending on the mineral. As with vitamins and other essential food nutrients, mineral requirements vary with animal species. For example, humans and other vertebrates need large amounts of calcium for construction and maintenance of bone and normal function of nerves and muscles. Phosphorus is an important constituent of adenosine triphosphate (ATP) and nucleic acid and is also essential for acid-base balance, bone and tooth formation. Red blood cells cannot function properly without iron in haemoglobin, the oxygen-carrying pigment of red blood cells. Iron is also an important component of the cytochromes that function in cellular respiration. Magnesium, copper, selenium, zinc, iron, manganese and molybdenum are important co-factors found in the structure of certain enzymes and are indispensable in numerous biochemical pathways. Vertebrates need iodine to make thyroid hormones. Sodium, potassium and chlorine are important in the maintenance of osmotic balance between cells and the interstitial fluid. Magnesium is an important component of chlorophyll in plants. The interactions between nutrition and diseases, nutrition and drug metabolism have been reported. Excessive intake of some minerals can upset homeostatic balance and cause toxic side effects. For example, excess sodium intake is associated with high blood pressure and excess iron can cause liver damage. Also, severe shortages or self-prescribed minerals can alter the delicate balance in body functions that promotes health. The knowledge of the biochemistry of the mineral elements is also essential because individuals suffering from a chronic illness or taking medications that affect the body's use of specific nutrients need to be enlightened. The aim of this paper is to review the biochemical functions and the importance of the mineral elements in health and disease conditions of humans, animals and plants as this will assist in the prevention of nutrition-related diseases and maintenance of good health for humans and animals that depend on plants for food. This paper could also serve as a ready source of literature review for researchers involved in nutritional sciences.

# INTRODUCTION

Minerals are inorganic substances, present in all body tissues and fluids and their presence is necessary for the maintenance of certain physicochemical processes which are essential to life. Minerals are chemical constituents used by the body in many ways. Although they yield no energy, they have important roles to play in many activities in the body (Malhotra, 1998; Eruvbetine, 2003). Every form of living matter requires these inorganic elements or minerals for their normal life processes (Hays and Swenson, 1985; Ozcan, 2003). Minerals may be broadly classified as macro (major) or micro (trace) elements. The third category is the ultra-trace elements. The macro-minerals include calcium, phosphorus, sodium and chloride, while the micro-elements include iron, copper, cobalt, potassium, magnesium, iodine, zinc, manganese, molybdenum, fluoride, chromium, selenium and sulfur (Eruvbetine, 2003). The macro-minerals are required in amounts greater than 100 mg/dl and the micro-minerals are required in amounts less than 100 mg/dl (Murray et al., 2000). The ultra-trace elements include boron, silicon, arsenic and nickel which have been found in animals and are believed to be essential for these animals. Evidence for requirements and essentialness of others like cadmium, lead, tin, lithium and vanadium is weak (Albion Research Notes, 1996). The mineral elements are separate entities from the other essential nutrients like proteins, fats, carbohydrates, and vitamins. Animal husbandry had demonstrated the need for minerals in the diet (Hegsted et al., 1976). In this century, biological assay methods clarified the significance and importance of mineral elements for human and animal nutrition and modern analytical techniques led to the detection of trace elements as essential nutrients and this is still an active area of current research. Micronutrient deficiencies are a major public health problem in many developing countries, with infants and pregnant women especially at risk (Batra and Seth, 2002). Infants deserve extra concern because they need adequate micronutrients to maintain normal growth and development (Rush, 2000). The micronutrient deficiencies which are of greatest public health significance are iron deficiency, causing varying degrees of impairment in cognitive performance, lowered work capacity, lowered immunity to infections, pregnancy complications e.g. babies with low birth weight, poor learning capacity and reduced psychomotor skills (Batra and Seth, 2002). Medical reports show that very severe anaemia is a direct cause of maternal and child mortality (Chakravarty and Ghosh, 2000). There have been suggestions that more than anything else, lack of adequate information about the composition of varied feed resources in some regions have been the major drawback to their utilization, rather than real shortage (Aletor and Omodara, 1994). For instance, there is very limited information on the mineral elements in some plants used as human food and animal feeds consumed in Nigeria, especially the newlyintroduced varieties of diets and the lesser known legumes. Some of the earlier information on mineral elements was based on analysis employing less sensitive methods, which may not be reliable. The aim of this review is to re-visit the provision of information on the importance of mineral elements to humans, animals and plants and also to emphasize on the need for their levels to be ascertained in water and commonly consumed plant foods using modern analytical techniques. This will assist Soetan et al. 201 greatly in providing relevant information on the importance of minerals to health and this would subsequently assist in prevention and management of mineral-associated deficiency diseases. Data on mineral contents of human foods and animal feeds are essential for formulation of feeding regimes and food processing techniques.

# THE IMPORTANCE OF THE MINERAL ELEMENTS

The importance of mineral elements in human, animal and plant nutrition has been well recognized (Underwood, 1971; Darby, 1976). Deficiencies or disturbances in the nutrition of an animal cause a variety of diseases and can arise in several ways (Gordon, 1977). When a trace element is deficient, a characteristic syndrome is produced which reflects the specific functions of the nutrient in the metabolism of the animal. The trace elements are essential components of enzyme systems. Simple or conditioned deficiencies of mineral elements therefore have profound effects on metabolism and tissue structure. To assess the dietary intake and adequacy of minerals, information needs to be collected on mineral element content of foods, diets and water (Rao and Rao, 1981; Simsek and Aykut, 2007). There is limited information on the trace element content of water and numerous plant foods consumed in some less developed countries. The significance of the mineral elements in humans, animals and plants nutrition cannot be overemphasized. The presence of mineral elements in animal feed is vital for the animal's metabolic processes. Grazing livestock from tropical countries often do not receive mineral supplementation except for common salt and must depend almost exclusively upon forage for their mineral requirements (McDowell et al., 1984). Mineral deficiencies or imbalances in soils and forages account partly for low animal production and reproductive problems. Soil acidity and season are factors affecting mineral uptake by plants. Plants use these minerals as structural components in carbohydrates and proteins; organic molecules in metabolism, such as magnesium in chlorophyll and phosphorus in ATP; enzyme activators like potassium, and for maintaining osmotic balance. Calcium is highly implicated in the maintenance of firmness of fruits (Olaiya, 2006) and its requirements in fruits are related to cell wall stability and membrane integrity (Belakbir et al., 1998). Mineral elements play important roles in health and disease states of humans and domestic animals. For example, iron deficiency anaemia and goitre due to iodine deficiency are reported to be problems of public health importance in some communities (Partwardhan, 1961; Deosthale and Belavady, 1978). Trace elements of significance to people with HIV are zinc and selenium. Selenium is an antioxidant that increases immune function. Zinc, usually taken to stimulate the immune system, has been reported to weaken immune system function and lower calcium evels in HIV – positive men (O' Connor, 1995; Wood, 2000).

## Methods of Analyses for the Minerals

Mineral elements in plant and soil samples are essentially estimated by colorimetric procedure according to the method of Sandell (1959). The wet-ashing procedure of Gorsuch (1959) using nitric acid, perchloric acid and sulphuric acid in the ratio of 3:2:1 is usually employed for the digestion of the food samples. Processing of samples of foods and water for mineral analysis are carried out in a dust-free room to avoid contamination, using distilled deionized water for all dilutions (Rao and Rao, 1981). Appropriate blanks are also included along with the samples and trace elements are estimated in the samples using an atomic absorption spectrophotometer. Internationally approved standards are also run along with the samples to be analysed to check the results obtained. Studying the chemical composition of plants provide clues about their nutritional requirements (Roberts, 1985). Researchers use a method known as hydroponic culture to determine which of the mineral elements are actually essential nutrients. In the technique known as hydroponic culture, a researcher baths the roots of plants in solutions of various minerals dissolved in known concentrations. Aerating the water will provide the roots with oxygen for cellular respiration. A particular mineral, such as potassium, can be ommited from the culture medium to test whether it is essential to the plants. If the element removed from the mineral solution is an essential nutrient, then the incomplete medium will cause plants to become abnormal in appearance compared with controls grown on a complete mineral medium. Analyses of the mineral content of a plant or soil are used to confirm the diagnosis of a specific mineral deficiency. Such studies have helped identify 17 elements that are essential nutrients in all plants and a few other elements that are essential to certain groups of plants as shown in Table 1.

# FACTORS AFFECTING THE CONCENTRATION AND AVAILABILITY OF THE MINERAL ELEMENTS

Several factors directly or indirectly influence the levels of minerals in plants and hence the amounts available for humans and animals that depend on plants for foods and feeds respectively. The amount of a particular nutrient in the diet may be insufficient to meet the requirements. However, the metabolism of the animal may be deranged by the interaction of dietary, environmental and genetic factors (Gordon, 1977). Location has been reported to influence the mineral and trace element compositions of rice, wheat, oats and barley (Basargin and Peregudora, 1969; Kavanek and Janicek, 1969) and these are mainly attributed to the altered soil conditions. The nature and chemical composition of the soil are also involved in the locational differences observed in the mineral elements present in grain sorghum (Deosthale and Belavady, 1978). Feeds grown on high-selenium soils are good sources of selenium and may be used in ration formulation for poultry in order to supply a source of selenium (Merck, 1986). Uptake of copper, zinc and manganese by plants is affected by the level of phosphate fertilizer (Mongia, 1966; Baser and Deo, 1967), while the chemical form of nitrogen fertilizer has been reported to affect the uptake of copper (Chand, 1969). Environmental factors such as location rather than genotype is reported to have greater influence on the mineral and trace element composition of sorghum grain (Belavady and Deosthale, 1978). Cobalt, copper, iodine and selenium deficiencies in the soil and flora in certain areas of the world have led to deficiencies of these minerals in domestic animals (Hays and Swenson, 1985). Also, the increment of selenium in the soil may lead to high levels in plants which are toxic to animals. Nutritional disorders involving the mineral elements may arise as simple deficiencies or excesses of particular elements but the extent to which the other organic or inorganic nutrients are present in the diet will determine the deficiencies or toxicities. The conditioning factors may be a reflection of the soils on which the plants are grown, or they may be related to the presence of specific plants that are seleniferous or goitogenic (Hays and Swenson, 1985). Mineral deficiencies or imbalances in soils or forages have been implicated, in part, for low animal production, and poor reproductive performance in the developing regions of the world (Aletor and Omodara, 1994). Antinutritional factors (ANF) present in plants could also affect the absorption and availability of some minerals by humans and animals. Anti-nutritional factors reduce the nutrient utilization and or food intake of plant foods (Osagie, 1998). The need for adequate processing to reduce the antinutritional factors in plants used as human foods and animal feeds have been reviewed by (Soetan and Oyewole, 2009). The levels of these substances in plants vary with the specie of plant, cultivar and post-harvest treatment. Examples of ANF which could reduce the bioavailability of minerals are oxalates and phytates. Oxalalic acid, like phytic acid, has the ability to bind some divalent metals such as calcium and magnesium thereby interfering with their metabolism (Blood and Radostits, 1989). Chelates bind many elements making them nutritionally unavailable, thereby inducing dietary deficiencies (Nelson et al., 1968). Zinc may be complexed with calcium-phytate and lead to inefficient utilization of dietary zinc. Adequate calcium intake will compensate for the reduced availability of calcium, but this will aggravate the zinc deficiency and such a relationship has frequently led to parakeratosis in pigs. Phytic acid reduces the absorption of calcium from the gastro-intestinal tract and consequently implicated in the development of rickets when chicks are fed cereals such as sorghum. Zinc and iron deficiency symptoms have been reported in man and chicken (Lease, 1966) when fed diets high in phytic acid. There is a significant inverse relationship between phytic acid content and the availability of calcium, magnesium, phosphorus and zinc in products like soya bean, palm kernel, rapeseed and cottonseed meals (Nwokolo and Bragg, 1977). Phytic acid may decrease the availability of divalent cations such as calcium, zinc and iron, by the formation of an insoluble protein-phytic acid- mineral complex, it has been cited as causing reduced availability of zinc in soybean foods (O'Dell and Savage, 1960; Forbes and Parker, 1977). Sometimes, either as a consequence of low availability of iodine from the soil, or the presence in the food of goitrogenic substances, which interfere with the utilization of iodine by the thyroid, the iodine requirements of some animals e.g. the sheep are not met in the natural diet or so iodine supplements must be fed (Merck, 1986).

**Table 1:** Functions and deficiency symptoms of macro- and micro- elements in plants

| Element | Form available to plants | Major functions | Deficiency symptoms |
|---|---|---|---|
| Macronutrients | | | |
| Carbon | $CO_2$ | Major component of plant's organic compounds | Growth retardation |
| Oxygen | $CO_2$ | Major component of plant's organic compounds | Plants become stunted |
| Hydrogen | $H_2O$ | Major component of plant's organic compounds | Plants become stunted |
| Nitrogen | $NO_3^-$, $NH_4^+$ | Component of nucleic acids, proteins, hormones, coenzymes and chlorophylls | Stunted growth, light green older leaves, older leaves become yellow and die (chlorosis) |
| Sulfur | $SO_4^{2-}$ | Component of proteins, two vitamins, coenzymes | Light green or yellow leaves, reduced growth |
| Phosphorus | $H_2PO_4^-$, $HPO_4^{2-}$ | Component of nucleic acids, phospholipids, adenosine triphosphate (ATP), several coenzymes | Purplish veins in older leaves, fewer seeds and fruits, stunted growth |

| | | | |
|---|---|---|---|
| Potassium | $K^+$ | Cofactor that functions in protein synthesis, activation of enzymes, major solute functioning in water balance and thus affecting osmosis, operation of stomata | Reduced growth, curled, mottled, or spotted older leaves, burned leaf margins, weakened roots and stems |
| Calcium | $Ca^{2+}$ | Important in formation and stability of cell walls and in maintenance of membrane structure and permeability, activates some enzymes, regulates many responses of cells to stimuli | Leaves are deformed, terminal buds die, reduced root growth |
| Magnesium | $Mg^{2+}$ | Component of chlorophylls, activates many enzymes | Chlorosis, drooped leaves |
| **Micronutrients** | | | |
| Chlorine | $Cl^-$ | Required for water-splitting step of photosynthesis, functions in water balance, role in root, shoot growth, plays a role in photolysis | Wilting, chlorosis, death of some leaves |

| | | | |
|---|---|---|---|
| Iron | $Fe^{3+}$, $Fe^{2+}$ | Component of cytochromes, electron transport, activates some enzymes, plays a role in chlorophyll synthesis | Chlorosis, yellow and green striping in grasses |
| Boron | $H_2BO_3$ | Cofactor in chlorophyll synthesis, may be involved in carbohydrate transport and nucleic acid synthesis, play roles in flowering, germination, fruiting, cell division, nitrogen metabolism | Terminal buds, lateral branches die, leaves thicken, curl, become brittle |
| Manganese | $Mn^{2+}$ | Active in formation of amino acids, activates some enzymes, coenzyme activity, required for water-splitting step of photosynthesis, chlorophyll synthesis | Light green leaves with green major veins, leaves whiten and fall off |

| | | | |
|---|---|---|---|
| Zinc | $Zn^{2+}$ | Active in formation of chlorophyll, activates some enzymes, plays a role in formation of auxin, chloroplasts, and starch | Chlorosis, mottled or bronzed leaves, abnormal roots |
| Copper | $Cu^{+}$,$Cu^{2+}$ | Component of many redox and lignin-biosynthetic enzymes | Chlorosis, dead spots in leaves, stunted growth, terminal buds die, necrosis in young leaves |
| Molybdenum | $MoO_4^{2-}$ | Essential for nitrogen fixation, cofactor that functions in nitrate reduction, component of enzyme used in nitrogen metabolism | Possible nitrogen deficiency, pale green, rolled or cupped leaves, mottling and necrosis in old leaves |
| Nickel | $Ni^{2+}$ | Cofactor for an enzyme functioning in nitrogen metabolism | |

Source: (Williams, 1992).

Note: All mineral elements contribute to the water-solute balance.

# INTERRELATIONSHIPS AND INTERFERENCES AMONG THE MINERAL ELEMENTS

There are many metabolic and absorptive interrelationships among the mineral elements which contribute to variations in the degree of physiological response to deficient or toxic levels.

These relationships make it difficult to determine the optimum dietary level for the individual elements required for humans and domestic animals. As a result of this, the recommended dietary level of any element should rarely be considered independent of the level of other essential nutrients (Hays and Swenson, 1985).

The functions of minerals in humans and animals are inter-related. Examples are the definite relationship of calcium and phosphorus in the formation of bones and teeth and as the major structural elements of the skeletal tissue.

Hypocalcaemia may cause weakness of the heart similar to that caused by hyperkalaemia. A high level of potassium appears to increase the requirement for sodium and *vice versa* (Merck, 1986). Potassium deficiency leads to an increase in the basic amino acid concentration of the tissue fluids and some increase in cellular sodium levels as a means of maintaining cation-anion balance. Potassium influences the contractibility of smooth, skeletal, and cardiac muscles and has an effect on muscular irritability that, like that of sodium, tends to antagonize the effect of the calcium ion.

Under conditions of salt restriction, calcium appears highly important in helping to maintain the potassium content of tissue (Hays and Swenson, 1985). The plasma phosphorus level is inversely related to the blood calcium level. There are also inter-relationships of iron, copper and cobalt (in vitamin B12) in haemoglobin synthesis and red blood cell formation (Hays and Swenson, 1985). It is reported that mercuric chloride toxicity in mammals can be overcome by co-administration of sodium selenite (Gailer et al., 2000). It has been repeatedly speculated that a deficiency in copper (Cu) or an excessive provision with dietary manganese (Mn) could play a role in spongiform encephalopathies [bovine (BSE) and others] (Clauss et al., 2007).

Copper functions in the utilization of iron in an early stage of haemopoiesis. Copper deficiency results in an increase in iron in the liver, whereas an excesss of copper results in a decrease in iron content of the liver, thus reflecting the role of copper in iron utilization. Copper is present in blood plasma as a copper-carrying plasma protein called erythrocuprin. Erythrocuprin provides a link between copper and iron metabolism and mediates the release of iron from ferritin and haemosiderin (Hays and Swenson, 1985). The dietary requirement of copper is affected by the level of some other minerals in the diet, and is increased in ruminants by excessive molybdenum. Treatment of copper poisoning is based on the rationale that excess molybdenum may cause copper deficiency and molybdenum in conjunction with the sulfate ion has been used in treating copper poisoning in ruminants (Pierson and Aenes, 1958). The Cu requirement varies among animal species to some extent but is influenced to a large degree by its relationship with and the intake of other mineral elements such as iron, molybdenum and sulfate.

Sodium, potassium, calcium, phosphorus and chlorine serve individually and collectively in the body fluids. Under stress conditions, a loss of sodium may be compensated for by an increase in potassium; but the animal is limited in its capacity to substitute bases, and major losses of sodium lead to a significant lowering of osmotic pressure, and therefore to a loss of water or dehydration. A close relationship exists between chloride and sodium ions. Chloride ion is accompanied by hydrogen ion in nearly equal amounts. In animals, excess chloride and a constant level of sodium can result in acidosis, whereas an excess sodium and a constant level of chloride can result in alkalosis. Intake of excess dietary manganese had been reported to interfere with phosphorus retention (Hays and Swenson, 1985). The amount of copper in the diet of animals necessary to prevent a copper deficiency is influenced by the intake of other dietary constituents, notably molybdenum and inorganic sulphate (Merck, 1986). High intake of molybdenum in the presence of adequate sulphate increases the requirement for copper in animals (sheep) (Merck, 1986).

Cardiac muscle, skeletal muscle and nervous tissue depend on a proper balance between calcium and magnesium ions. A high calcium intake will increase the need for zinc. The symptoms of magnesium deficiency resemble that of low-calcium tetany (Hays and Swenson, 1985). Deficiencies of trace elements like zinc, copper and magnesium have been implicated in various reproductive events like infertility, pregnancy wastage, congenital anomalies, pregnancy-induced hypertension, Soetan et al. 205 placental abruption, premature rupture of membranes, still births and low birth weight (Pathak and Kapil, 2004). An excess of dietary fat or poor digestion of fat may reduce calcium absorption through the formation of insoluble calcium soaps; however, small amounts of fat may improve calcium absorption. An excess of iron, aluminium or magnesium interferes with phosphorus absorption through the formation of insoluble phosphates. In plants, a high level of fertilization results in high levels of nitrogen and potassium and this tends to reduce the uptake of magnesium by the plant. In turn, the high level of potassium interferes with the absorption of magnesium, and the high levels of dietary nitrogen provided by the plants increase the excretion of magnesium. These effects lead to a physiological deficiency of magnesium and can result in tetany and death (Hays and Swenson, 1985). A reciprocal antagonism exists between molybdenum and copper. Molybdenum, in the presence of sulfates, limits copper retention in cattle and sheep; however, neither molybdenum or sulfates alone affect copper retention. Chronic copper poisoning associated with high levels of copper in the liver of ruminants had been observed under conditions of moderate intake of copper but very low dietary levels of molybdenum and sulfate.

Conversely, excessive levels of molybdenum and sulfate may cause characteristic symptoms of copper deficiency even with apparently adequate intake of copper (Hays and Swenson, 1985). Trace elements and anti-oxidants have also been shown to influence host cellular and humoral immunological functions. The immune system utilizes these essential minerals and factors to meet the demands of challenges by infectious agents (Spallholz et al., 1990; Sherman, 1990). Trace elements are essential in the maturation, activation and functions of host defence mechanism (Bendich, 1990; Romero, 2003). Cu, Fe, Se, Mn and Zn are integral part of enzymatic anti-oxidants. Arinola et al. (2008b) reported that infection with malaria parasites during pregnancy leads to raised plasma trace elements and total antioxidants as a result of free radical generation. Trace metals are known to play important roles in the catalytic activities of major antioxidant enzymes (Arinola et al., 2008a). Trace elements are required in small concentrations as essential components of biological enzyme systems or of structural portions of biologically active constituents (Arinola et al., 2008c).

Immune cells, like all other types of cells, require an adequate supply of trace elements (Fe, Cu and Zn) to express and preserve the structure and function of key metalloproteins that participate in house keeping processes such as energy production and to protect the cell against highly toxic reactive oxygen species. Also adequate level of Fe and Zn is required for continous generation of immune cells in bone marrow and the clonal expansion of lymphocytes in response to antigenic stimulation (Chandra, 1990). Neurologic disorders and trace elements are closelyrelated (Arinola et al., 2008c). Non-living plant biomass have the ability to bind metals and this has been attributed to the presence of various functional groups, which can attract and sequester the metal ions (Horsfall et al., 2003). Cu and Zn are an integral part of Cu-ZnSOD (superoxide dismutase). Fe is an integral part of catalase, Mn is an integral part of Mn-SOD while Se is an integral part of Se-GPX (glutathione peroxidase) which is a major scavenger of H2O2 (Arinola et al., 2008a). The SOD catalyses the dismutation of of superoxide to H2O2, which must be removed by catalase or glutathione peroxidase. Deficiency of Se, Zn and Cu decreases activities of superoxide dismutase and glutathione peroxidase (El- Behairy et al., 1997; Ellis and Salt, 2003). It has been reported that plants can use rich source of nitrogen, phosphorus and potassium for growth if suitably managed (Sharma et al., 2008). However, some elements serve very specific roles such as iodine in thyroxine and cobalt as an integral part of vitamin B12. Thyroxin and vitamin B12 are also intimately involved in processes related to many other organic and inorganic nutrients. Strontium and calcium have a similar and interrelated physiological behaviour. Rubidium is found in animal tissue and it resembles potassium in its distribution and excretory pattern (Hays and Swenson, 1985). Thc amelioration of mercury toxicity by selenium shows the importance of interactions of selenium with other elements (Panzek and Ostadalova, 1967). Studies of these interactions on a molecular level indicate a profound effect of selenium on mercury metabolism in animals (Burk et al., 1974). Arsenic has long been known to decrease selenium toxicity under some conditions. Increased biliary excretion of selenium accompanies arsenic administration and is the presumed mechanism (Lexander and Baumann, 1966). Interactions of selenium with several other elements are known and new developments in this area are being expected.

# THE BIOCHEMISTRY AND FUNCTIONS OF THE INDIVIDUAL MINERAL ELEMENTS

The basic functions performed by the minerals are: they are structural components of body tissues, are involved in the maintenance of acid-base balance and in the regulation of body fluids, in transport of gases and in muscle contractions (Malhotra, 1998; Murray et al.,2000).

Knowledge of the biochemistry and functions of the mineral elements in humans, animals and plants will assist plant physiologists and breeders/geneticists, to give priority to mineral elements of importance in health and disease of humans and animals when selecting desirable traits in diets and this will also assist food industries, dieticians, human and animal nutritionists, and veterinarians to be aware of the effects of different processing methods/techniques on these important mineral elements . For example, high dietary sodium is implicated in cardiovascular and renal disorders. Consequently, high dietary sodium is often discouraged in patients/subjects who suffer from or are prone to hypertension. Also, knowledge of the importance of the mineral elements in plants is essential as the global trend in nutrition and medicine is shifting towards the consumption of plant foods (fruits and vegetables) and medicinal plants (phyto-medicines) respectively, because the plant kingdom is reported to be full of large numbers of beneficial substances to both human and animal health (Igile, 1995; Nwadiaro and Nwachukwu, 2007; Ogbonna et al., 2007; Soetan, 2008).

The symptoms of a mineral deficiency depend on the function and mobility of the element. The metabolism of soil bacteria makes nitrogen available to plants (Roberts, 1985). Carnivorous plants supplement their mineral nutrition by digesting animals (Roberts, 1985). Plants need sunlight as the energy source for photosynthesis, but to synthesize organic matter, plants also require raw materials in the form of inorganic substances: carbon dioxide, water, and a variety of minerals present as inorganic ions in the soil. With its ramifying root system and shoot system, a plant is extensively networked with its environment-soil and air, which are the reservoirs of the plant's inorganic nutrients (Roberts, 1985). A particular chemical element is considered to be an essential nutrient if it is required for a plant to grow from a seed and complete the life cycle, producing another generation of seeds.

## Biochemical Functions of Mineral Elements in Humans And Animals

### *Calcium (Ca)*

Calcium functions as a constituent of bones and teeth, regulation of nerve and muscle function. In blood coagulation, calcium activates the conversion of prothrombin to thrombin and also takes part in milk clotting. It plays a vital role in enzyme activation. Calcium activates large number of enzymes such as adenosine triphosphatase (ATPase), succinic dehydrogenase, lipase etc. It is also required for membrane permeability, involved in muscle contraction, normal transmission of nerve impulses and in neuromuscular excitability. A reduced extracellular blood calcium increases the irritability of nerve tissue, and very low levels may cause spontaneous discharges of nerve impulses leading to tetany and convulsions (Hays and Swenson, 1985; Malhotra, 1998; Murray et al., 2000). Calcium absorption requires calcium-binding proteins and is regulated by vitamin D, sunlight, parathyroid hormone and thyrocalcitonin. Thyrocalcitonin decreases plasma calcium and phosphate levels whereas parathyroid hormone increases them. Dietary calcium and phosphorus are absorbed mainly in the upper small intestine, particularly the duodenum and the amount absorbed is dependent on source, calcium-phosphorus ratio, intestinal pH, lactose intake and dietary levels of calcium, phosphorus, vitamin D, iron, aluminium, manganese and fat. The greater the need, the more efficient the absorption. Absorption of calcium and phosphorus is facilitated by a low intestinal pH which is necessary for their solubility and thus normal gastric secretion of hydrochloric acid or H+ is necessary for efficient absorption. Achlorhydria decreases absorption of these minerals. The low pH of the duodenum accounts for the greater absorption in that area. Lactose also enhances the aborption of calcium (Hays and Swenson, 1985).

In plants, calcium is taken up in the ionized form (as $Ca^{2+}$), the leafy parts are relatively high in calcium and low in phosphorus, whereas, the reverse is true of the seeds. Legumes, in general, have higher calcium content than grasses (Merck, 1986). In children, calcium deficiency causes rickets due to insufficient calcification by calcium phosphate of the bones in growing children. The bones therefore remain soft and deformed by the body weight. In adults, it causes osteomalacia, a generalized demineralization of bones. It may also contribute to osteoporosis, a metabolic disorder resulting in decalcification of bone with a high incidence of fracture, that is, a condition where calcium is withdrawn from the bones and the bones become weak and porous and then breaks (Hays and Swenson, 1985; Malhotra, 1998; Murray et al., 2000). Calcium deficiency also affects the dentition of both children and adult. Toxicity symptoms occur with excess absorption due to hypervitaminosis D or hypercalcaemia due to hyperparathyroidism, or idiopathic hypercalcaemia. Excess calcium depresses cardiac activity and leads to respiratory and cardiac failure; it may cause the heart to stop in systole, although, normally, calcium ions increase the strength and duration of cardiac muscle contraction. Excess calcium and phosphorus are excreted by the kidney. Ca and P excreted in faeces are largely the unabsorbed dietary minerals; some comes from the digestive juices, including bile (Hays and Swenson, 1985). Growing, pregnant and especially lactating humans and animals require liberal amounts of calcium and phosphorus. Parturient paresis, or milk fever, in cows is associated with calcium metabolism. This illness usually occurs with the onset of profuse lactation and the most common abnormality is acute hypocalcaemia with decline in blood calcium level from normal. Serum magnesium levels may be elevated or depressed, low levels being accompanied by tetany and high levels by a flaccid paralysis. Sources of calcium include Beans, lentils, nuts, leafy vegetables, dairy products, small fishes including sardines, bones, etc.

## *Phosphorus (P)*

Phosphorus is located in every cell of the body and is vitally concerned with many metabolic processes, including those involving the buffers in body fluids (Hays and Swenson, 1985). It functions as a constituent of Soetan et al. 207 bones, teeth, adenosine triphosphate (ATP), phosphorylated metabolic intermediates and nucleic acids. It serves buffering action, that is, phosphate buffers, functions in the formation of high energy compounds, that is, adenosine triphosphate (ATP) and is involved in the synthesis of phospholipids and phosphoproteins. Practically, every form of energy exchange inside living cells involve the forming or breaking of high-energy bonds that link oxides of phosphorus to carbon or to carbon-nitrogen com-pounds (Hays and Swenson, 1985; Malhotra, 1998; Murray et al., 2000). Vitamin D is probably involved in the control of phosphorus absorption and serum levels are regulated by kidney reabsorption. Phosphorus is an essential macronutrient for plants and one of the three nutrients generally added to soils in fertilizers because of its vital role of energy transfer in living organisms and in plants. Adequate phosphorus availability stimulates early growth and hastens maturity in plants (Sharma et al., 2008).

Phosphorus is also needed for soil fertility. In plants, as grasses mature, phosphorus is transferred to the grain. Also, the phosphorus content of the plant is influenced markedly by the availability of phosphorus in the soil. As a result of this, low-quality pastures devoid of legumes and range plants tend to be naturally low in phosphorus, as the forage matures and the seeds fall; characteristically, the range soil is also deficient in phosphorus (Merck, 1986). A large percentage (60-80%) of the total phosphorus of cereal grains and oil seeds exists organically bound as phytic acid. Phytic acid, the hexaphosphoric acid ester of inositol, is present in cereal and legume seeds primarily as the Ca-Mg salt called phytin. The organically bound phosphorus, phytin phosphorus, is largely unavailable to monogastric animals, whereas ruminants can utilize it relatively very well. This species difference is explained by the presence of the enzyme phytase from rumen microorganisms, which hydrolyzes the organically bound phosphorus and renders it available for absorption. This partly, coupled with the slower growth rate of ruminants, accounts for the rather large difference in phosphorus requirements of ruminant and nonruminant animals (Hays and Swenson, 1985).

Decrease in serum phosphorus is found in rickets, hyperparathyroidism, De Toni-Fanconi Syndrome. Deficiency disease or symptoms in children causes rickets and in adults, it causes osteomalacia. Increase in serum phosphorus is found in chronic nephritis and hypoparathyroidism. Toxicity disease or symptoms include low serum $Ca^{2+}$ : P ratio. It may also lead to bone loss (Malhotra, 1998; Murray et al., 2000). Sources of phosphorus include phosphate food additives, green leafy vegetables and fruits, especially banana.

## Sodium (Na)

Sodium is the principal cation in extracellular fluids. Itregulates plasma volume and acid-base balance, involved in the maintenance of osmotic pressure of the body fluids, preserves normal irritability of muscles and cell permeability, activates nerve and muscle function and involved in $Na^+/K^+$-ATPase, maintenance of membrane potentials, transmission of nerve impulses and the absorptive processes of monosaccharides, amino acids, pyrimidines, and bile salts. The changes in osmotic pressure are largely dependent on sodium concentration (Hays and Swenson, 1985; Malhotra, 1998; Murray et al., 2000). Its metabolism is regulated by aldosterone.

Commonly used vegetable foodstuffs do not contain sufficient quantities of sodium to meet the animal's dietary need. This inadequacy is compensated for by including sodium chloride, common salt, in their diet or by allowing them to consume salt *ad libidum*. Sodium is readily absorbed as the sodium ion and circulates throughout the body. Excretion occurs mainly through the kidney as sodium chloride or phosphate. There are appreciable losses in perspiration, and the quantities lost by this route vary rather markedly with the environmental humidity (Hays and Swenson, 1985). Sodium deficiency in young chicks cause growth retardation. Egg production and hatchability in laying chickens are depressed (Merck, 1986).

Increased level of sodium in the serum is called hypernatraemia and this occurs in Cushion's disease, administration of adrenocorticotropic hormone (ACTH), administration of sex hormones, diabetes insipidous and after active sweating (Malhotra, 1998). Excessive intake of sodium chloride may result in salt toxicity which is mainly caused by sodium ion, since sodium acetate or sodium propionate affects the animals in a manner similar to that of sodium chloride. The amounts required for toxicity vary and are largely dependent on the availability of water to the animals. Toxicity usually occurs when animals are deprived of salt and then have access to a brine solution or loose salt without access to sufficient water (Hays and Swenson, 1985). Low level of sodium in the serum is hyponatraemia and this occurs in acute Addison's disease, vomiting, diarrhea, nephrosissevere burns and intestinal obstruction (Malhotra, 1998). Toxicity disease or symptoms may cause hypertension in susceptible individuals. Sources include table salt, salt added to prepared foods and most natural foods contain sodium.

## *Potassium (K)*

Potassium is the principal cation in intracellular fluid and functions in acid-base balance, regulation of osmotic pressure, conduction of nerve impulse, muscle contraction particularly the cardiac muscle, cell membrane function and $Na^+/K^+$-ATPase. Potassium is also required during glycogenesis. It also helps in the transfer of phosphate from ATP to pyruvic acid and probably has a role in many other basic cellular enzymatic reactions. Its metabolism is regulated by aldosterone. Hyperkalaemia is increased level in serum potassium and this occurs in Addison's disease, advanced chronic renal failure, shock and dehydration. Toxicity disease or symptoms include dilatation of the heart, cardiac arrest, small bowel ulcers. Hypokalaemia is low level of serum potassium and this occurs in diarrhoea, metabolic alkalosis and familial periodic paralysis. When lactating dairy cows have hypokalaemia, the milk production is markedly lowered.

Deficiency disease or symptoms occurs secondary to illness, functional and structural abnormalities including impaired neuromuscular functions of skeletal, smooth, and cardiac muscle, muscular weakness, paralysis, mental confusion (Hays and Swenson, 1985; Malhotra, 1998; Murray et al., 2000). Others are cardiac arrythmias, impaired carbohydrate tolerance, altered electrocardiogram in calves. Potassium deficiency affects the collecting tubules of the kidney, resulting in the inability to concentrate urine, and also causes alterations of gastric secretions and intestinal motility (Streeten and Williams, 1952). The rapidly growing animals apparently have a higher requirement for potassium, and increasing the protein level increases the requirement. Plant products contain many times as much potassium as sodium. Sources include vegetables, fruits, nuts.

## *Chlorine (Cl)*

Chlorine is involved in fluid and electrolyte balance, gastric fluid and chloride shift in $HCO_3$ - transport in erythrocytes. Chloride is the principal anion in extracellular fluid. It is involved in the regulation of extracellular osmotic pressure and makes up over 60% of the anions in this fluid compartment and is thus important in acidbase balance. The concentration of chloride ion is subject to more variation than that of sodium, since other anions, especially bicarbonates, can exchange for the chloride. It is the chief anion of the gastric juice and is accompanied by the hydrogen ions in nearly equal amounts. The chloride of the gastric secretions is derived from blood chloride and is normally reabsorbed during the latter stages of digestion in the lower intestine (Hays and Swenson, 1985; Murray et al., 2000).

It is the major negative ion of the extracellular fluid and participates in acid production in the stomach. Deficiency disease or symptoms occur in infants fed salt-free formula. On a chloride-deficient diet, the excretion of chloride in the urine or perspiration is markedly reduced (Hays and Swenson, 1985; Murray et al., 2000). It is also secondary to vomiting, diuretic therapy, renal disease. Excessive depletion of chloride ions through losses in the gastric secretions or by deficiencies in the diet may lead to alkalosis due to an excess of bicarbonate, since the inadequate level of chloride is partially compensated for or replaced by bicarbonate. Chloride is excreted in the faeces, sweat, and urine primarily as sodium orpotassium chloride, although it may be accompanied by ammonium ions when base needs to be conserved (Hays and Swenson, 1985). Sources include table salt and drinking water.

## *Magnesium (Mg)*

Magnesium is an active component of several enzyme systems in which thymine pyrophosphate is a cofactor. Oxidative phosphorylation is greatly reduced in the absence of magnesium. Mg is also an essential activator for the phosphate-transferring enzymes myokinase, diphophopyridinenucleotide kinase, and creatine kinase. It also activates pyruvic acid carboxylase, pyruvic acid oxidase, and the condensing enzyme for the reactions in the citric acid cycle. It is also a constituent of bones, teeth, enzyme cofactor, (kinases, etc) (Murray et al., (2000). The health status of the digestive system and the kidneys significantly influence magnesium status. Magnesium is absorbed in the intestines and then transported through the blood to cells and tissues. Approximately one-third to one-half of dietary magnesium is absorbed into the body. Gastrointestinal disorders that impair absorption such as Crohn's disease can limit the body's ability to absorb magnesium. These disorders can deplete the body's stores of magnesium and in extreme cases may result in magnesium deficiency.

When a magnesium-deficient diet is fed to young chicks, it leads to poor growth and feathering, decreased muscle tone, ataxia, progressive incoordination and convulsions followed by death (Merck, 1986).

Chronic or excessive vomiting and diarrhea may also result in magnesium depletion. Deficiency diseases or symptoms is secondary to malabsorption or diarrhoea, alcoholism. Acute magnesium deficiency results in vasodilation, with erythemia and hyperaemia appearing a few days on the deficient diet. Neuromuscular hyperirritability increases with the continuation of the deficiency, and may be followed eventually by cardiac arrhythmia and generalized tremours. A common form of magenesium-deficiency tetany in ruminants is called grass tetany or wheat wheat-pasture poisoning. This condition occurs in ruminants grazing on rapidly growing young grasses or cereal crops and develops very quickly. The physiological deficiency of magnesium can be prevented by magnesium supplementation of a salt or grain mixture and adequate consumption is also very important (Hays and Swenson, 1985). Toxicity disease or symptoms of magnesium deficiency in humans include depressed deep tendon reflexes and respiration (Murray et al., 2000). Sources include leafy green vegetables (containing chlorophyll).

## *Chromium (Cr)*

Chromium is an essential element for animals and Soetan et al. 209 humans (Frieden, 1984). It has been found in nucleoproteins isolated from beef liver and also in RNA preparations (Uppala et al., 2005). It could play a role in maintaining the configuration of the RNA molecule, because Cr has been shown to be particularly effective as a cross-linking agent for collagen (Eastmond et al., 2008). Cr has also been identified as the active ingredient of the glucose tolerant factor (Brown, 2003), a dietary factor required to maintain normal glucose tolerance in the rat. Trivalent chromium is a constituent of "glucose tolerance factor" (GTF), which binds to and activates/potentiates insulin action (Wennberg, 1994; Murray et al., 2000).

Chromium compounds have a wide variety of industrial uses, including production of stainless steel and other alloys, high-melting refractory materials, pigments and mordants for paints and dyes, tanned leather goods and chrome plating (Frieden, 1984). Cr affects the action of insulin in protein metabolism, as indicated by rats fed chromium-deficient diets repleted by chromium (Roginski and Mertz, 1969). Insulin-mediated amino acid transport into tissues was enhanced and incorporation of labeled glycine, serine and methionine into heart protein was greater in chromium-supplemented animals. Evidence of a role for chromium in lipid metabolism and chromium deficiency in the development of atherosclerosis is accumulating from animal and human studies (Frieden, 1984). Cr deficiencies may exist, particularly in children suffering from protein-calorie malnutrition (Mertz, 1974). In experimental animals, Cr deficiency leads to a reduced rate of removal of ingested glucose, due to a low sensitivity of peripheral tissues to insulin. Chromium is needed for growth of rats and its deficiency leads to a reduced life span, corneal lesions and interference with insulin action producing a diabetic state and this causes removal of glucose from the blood at a rate that is onehalf that of animals on a chromium-containing diet (Wennberg, 1994; Juturu and Komorowski, 2003). The Cr content of foodstuffs varies widely and is present in combination with a small organic molecule, the glucose tolerant factor. Chromium poisoning in humans is usually limited to accidental ingestion of chromic acid or chromates. Toxicity to kidney, liver, nervous system and blood are the major causes of death (Langard, 1980). Sources include meat, liver, brewer's yeast, whole grains, nuts, cheese.

## Cobalt (Co)

Cobalt is required as a constituent of vitamin B12 and its metabolism is the same as for vitamin B12. In addition to its role in vitamin B12, cobalt is also a cofactor of enzymes involved in DNA biosynthesis and amino acid metabolism (Arinola et al., 2008c). In cattle and sheep, bacteria in the rumen can use metallic Co to synthesize vitamin B12 and are thus the ultimate source of the vitamin in human diets. Vitamin B12 also plays a role in methylating cholineand thamine. The latter is required for the synthesis of DNA, which regulates cell division and growth. Co is readily absorbed into the bloodstream and excreted primarily in the urine. Deficiency disease or symptoms is manifested in vitamin B12 deficiency. Co deficiency in ruminants have been successfully alleviated by the use of cobalt oxide pellets, which remain in the reticulum or rumen fluid (Hays and Swenson, 1985). In humans, toxicity disease or symptoms include goitre, hypothyroidism and heart failure (Murray et al., 2000). In animals, excessive intake results in polycythaemia, apparently due to the inhibition by cobalt of certain respiratory enzyme systems, for example, cytochrome oxidase and succinic dehydrogenase. Deficiencies of cobalt in ruminants cause anorexia, wasting of skeletal muscle, fatty livers, haemosiderosis of the spleen and anaemia (Hays and Swenson, 1985). Dietary sources of cobalt are the same as for vitamin B12, such as foods of animal origin or fermented foods where the bacteria produce the vitamin. Organ meats are the best source of vitamin B12 (liver, kidney, heart, and pancreas), followed by clams, oysters, extra-lean beef, seafood, eggs, milk and yogurt, chicken, cheese, and miso (a fermented soybean product).

## Copper (Cu)

Copper is a constituent of enzymes like cytochrome c oxidase, amine oxidase, catalase, peroxidase, ascorbic acid oxidase, cytochrome oxidase, plasma monoamine oxidase, erythrocuprin (ceruloplasmin), lactase, uricase, tyrosinase, cytosolic superoxide dismutase etc. and it plays a role in iron absorption (Chandra, 1990). Cu is an essential micro-nutrient necessary for the haematologic and neurologic systems (Tan et al., 2006). It is necessary for the growth and formation of bone, formation of myelin sheaths in the nervous systems, helps in the incorporation of iron in haemoglobin, assists in the absorption of iron from the gastrointestinal tract (GIT) and in the transfer of iron from tissues to the plasma (Malhotra, 1998; Murray et al., 2000).

It is transported by albumin; bound to ceruloplasmin. Ceruloplasmin has oxidase activity and thereby facilitates the incorporation of ferric iron into transferrin. The copper-containing protein in red blood cells (rbc) is erythrocuperin, in liver, it is hepatocuperin and in brain, it is cerebrocuperin. In the monogastric animals, copper is absorbed mainly in the upper part of the small intestine, where the pH of the contents is still acidic. In general, Cu is poorly absorbed, and under normal conditions >90% of the ingested copper appears in the faeces. Most of the faecal copper is unabsorbed dietary copper, but some of it comes from the bile, which is the major pathway of Cu excretion. Biliary obstruction increases the excretion of copper through the kidney and intestinal wall (Hays and Swenson, 1985). Increased levels of copper are seen in acute infections and in chronic conditions such as cirrhosis, rheumatoid arthritis and in post-operative stages. Increased level is also found in malnutrition (Malhotra, 1998). Clinical disorders associated with Cu deficiencies include anaemia, bone disorders, neonatal ataxia, depigmentation and abnormal growth of hair, fur or wool, impaired growth and reproductive performance, heart failure and gastrointestinal disturbances. The incidence of these disorders varies widely among animal species. Cu deficiency has also been associated with cardiac hypertrophy and sudden cardiac failure. Gardea- Torresdey et al. (1990) reported that carboxyl groups found on the cell walls of dead algal biomass are potentially responsible for copper binding. Toxicity disease or symptoms are rare and is secondary to Wilson's disease (Murray et al., 2000).

In Wilson's disease, a large amount of copper is deposited in liver, brain, etc. Total copper content in the plasma and ceruloplasmin-bound copper content decreases and there is an increased excretion of copper in the urine. Sometimes, Cu may be deposited in the renal tubules giving rise to renal tubular degeneration and this is manifested as glycosuria and amino aciduria (Malhotra, 1998). Excess dietary Cu causes an accumulation of Cu in the liver with a decrease in blood haemoglobin concentration and packed cell volume. Liver function is adversely affected in copper poisoning. Jaundice results from erythrocyte haemolysis and this may lead to death unless treatment is started. In animals, sheep are more susceptible than cattle to the toxic effects of copper (Merck, 1986). Deficiency disease or symptom include anaemia (hypochromic, microcytic). Sources include liver, whole grains, molasses, legumes, nuts, shell fish and other seafoods.

## Iodine (I)

Iodine is a basic component of the thyroid hormones, thyroxine and mono-, di-, and tri-iodothyronine and it is stored in thyroid as thyroglobulin (Hays and Swenson, 1985; Murray et al., 2000). A few cases of goiter, or enlarged thyroids, have been observed in poultry, probably because it is difficult to observe the glands. Goitre has been produced experimentally in chickens by feeding a diet exceedingly low (about 0.025 ppm) in iodine to laying hens. Iodine ion is freely diffusible, is readily absorbed from the gastrointestinal tract, and is excreted mainly in the urine at a relatively constant rate provided the dietary intake is sufficient. Some older varieties of rapeseed meal, which were high in goitrogenic factors, invariably resulted in birds with enlarged thyroids if large quantities were used in the ration. Goitrogens have been demonstrated in the milk of cows following their ingestion of cruciferous plants in grazing or as fodder. Iodine deficiency was reported to be the most widespread of all mineral deficiencies in grazing livestock. A deficiency of iodine can be prevented by feeding stabilized iodized salt to animals or in poultryby supplementing the feed with as little as 0.35 mg/kg of iodine (Merck, 1986). Plants are highly variable in iodine content depending on the species, soil type including its iodine content, fertilizer, and climate. In children, deficiency causes cretinism and in adults, goiter and hypothyroidism and myxedema. The toxicity diseases or symptoms of iodine are thyrotoxicosis, goiter. Sources include iodized salt, sea food.

## *Iron (Fe)*

Iron functions as haemoglobin in the transport of oxygen. In cellular respiration, it functions as essential component of enzymes involved in biological oxidation such as cytochromes c, c1, a1, etc (Malhotra, 1998). Fe is an important constituent of succinate dehydrogenase as well as a part of the haeme of haemoglobin (Hb), myoglobin and the cytochromes (Chandra, 1990). Iron is required for proper myelination of spinal cord and white matter of cerebellar folds in brain and is a cofactor for a number of enzymes involved in neurotransmitter synthesis (Larkin and Rao, 1990). Iron is involved in synthesis and packaging of neurotransmitters, their uptake and degradation into other iron-containing proteins which may directly or indirectly alter brain function (Beard, 2001).

Iron exists in the blood mainly as haemoglobin in the erythrocytes and as transferrin in the plasma. It is transported as transferrin; stored as ferritin or haemosiderin and it is lost in sloughed cells and by bleeding (Murray et al., 2000). Fe is required for making Hb and it is a prooxidant which is also needed by microorganisms for proliferation (Galan et al., 2005). Biologically important compounds of iron are haemoglobin, myoglobin, cytochromes, catalases and peroxidase (Malhotra, 1998). Factors effecting the absorption of iron are: low phosphate diet which increases iron absorption, whereas high phosphate diet decreases iron absorption by forming insoluble iron phosphates. Adrenocortical hormones (glucocorticoids) play a role in regulating the level of plasma iron. During stress, when the hypothalamus, adenohypophysis, and adrenal cortex are activated, regardless of the source, the plasma iron decreases (Hays and Swenson, 1985). Iron in ferrous form is more soluble and is readily absorbed than the ferric form. Phytic acid and oxalic acid decreases iron absorption by forming iron phytate and iron oxalate. The absorption of iron is inhibited by profuse diarrhoea, malabsorption syndrome, achlorohydria, dissertion of small intestine and partial or total gastrectomy (Malhotra, 1998).

The plasma iron content is determined by the extent of blood losses, role of erythropoeisis, rate of apoferritin synthesis, rate of iron absorption from intestines and rate of red blood cell destruction. Deficiency disease or symptoms include anaemia, (hypochromic, microcytic). Fe deficiency has been reported to have a role in brain development and in the pathophysiology of restless legs Soetan et al. 211 syndrome (Tan et al., 2006). Also, Fe deficiency is associated with alterations in many metabolic processes that may impact brain functioning, among whom are neurotransmitter metabolism, protein synthesis, organogenesis etc (Beard, 1999). Early iron deficiency has also been reported to affect GABA metabolism in adult rats (Youdim et al., 1989; Taneja et al., 1986). Fe accumulation has been related to some neurologic disorders such as Alzheimer disease, Parkinson disease, type-1 neuro-degeneration with brain iron accumulation and other disorders (Sadrzadeh and Saffari, 2004). Brain is quite sensitive to dietary iron depletion and uses a host of mechanisms to regulate iron flux homostatically (Batra and Seth, 2002). The pig is born with low iron stores and develops an iron deficiency anaemia if not provided with supplementary iron. The factors causing the onset of anaemia in piglets are its relatively low iron stores at birth, its high growth rate early in life, and the low level of iron in sow milk. If the pig is given iron supplements at birth, the total red cell mass or volume per unit of body weight increases from birth to three weeks of age (Hays and Swenson, 1985).

Iron deficiency anaemia also occurs at birth in other animals like dogs, cats, cattle, etc but it is more pronounced in pigs. Excessive accumulation of iron in the liver, pancreas, heart, lungs and other tissues cause haemosiderosis and when this is accompanied by bronze pigmentation of the skin, the condition is called haemochromatosis (Malhotra, 1998; Murray et al., 2000). Sources include red meat, spleen, heart, liver, kidney, fish, egg yolk, nuts, legumes, molasses, iron cooking ware, dark green leafy vegetables.

## *Manganese (Mn)*

Manganese is a cofactor of hydrolase, decarboxylase, and transferase enzymes (Murray et al., 2000). It is involved in glycoprotein and proteoglycan synthesis and is a component of mitochondrial superoxide dismutase. Manganese is a co-factor in phosphohydrolases and phosphotransferases involved in the synthesis of proteoglycans in cartilage. Mn is a part of enzymes involved in urea formation, pyruvate metabolism and the galactotransferase of connective tissue biosynthesis (Chandra, 1990). Mn activates several important enzyme systems and in this capacity it is required for the synthesis of acid mucopolysaccharides, such as chondroitin sulphate, to form the matrices of bones and egg shells. Consequently skeletal deformities and defects in shell quality occur when the manganese intake is inadequate (Gordon, 1977). The fact that Mn is concentrated in the mitochondria has led to the suggestion that, *in vivo*, manganese is involved in the partial regulation of oxidative phosphorylation. Absorption of Mn is inhibited by the presence of excessive amounts of calcium and phosphorus in the diet.

The absorption and retention of manganese from foodslow in iron, such as milk, are relatively high. If milk is supplemented with iron, the percentage of manganese absorbed is reduced (Gruden, 1977). Increased absorption of manganese has been reported during pregnancy in sows (Kirchgessner et al., 1981) and with coccidiosis infection in chickens (Southern and Baker, 1983). Mn deficiency has been demonstrated in several animal species including laboratory animals, pigs, poultry, and possibly in cattle. Its severity depends greatly on the degree and duration of the deficiency and on the maturity of the animal (Hays and Swenson, 1985). Manganese deficiency presents with the following signs; in pigs, lameness, enlarged hock joints, and shortened legs, in cattle, leg deformities with overknuckling, in chicks, poults and ducklings, perosis or slipped tendon; and in chick embryos, nutritional chondrodystrophy. In laboratory animals, effects of deficiency include deformities of bone, poor growth, impaired reproduction, egg shell formation, and blood clotting. Some of these defects are related to the role of the manganese ion as the most effective activator of glycosyl transferase enzymes in the synthesis of mucopolysaccharides and glycoproteins (Leach, 1974).

Other deficiency disease symptoms are ataxia and abnormal formation of otoliths in the inner hear. In other species, congenital defects in embryonic bone development result from Mn deficiency. Birds are much more susceptible to manganese deficiency than mammals because their requirements for this element are considerably higher and this is attributable partly to relatively poor absorption from the intestine (Gordon, 1977). Deficiency disease or symptoms is unknown in humans. Mn overexposure reportedly may have an adverse effect on central nervous system (CNS) function and mood (Tan et al., 2006). Toxicity disease or symptoms by inhalation poisoning produces psychotic symptoms and parkinsonism. Corn is extremely low in manganese (4-12 ppm) and so animals fed high-corn diets especially if supplemented with animal by-products, which are also low in manganese content, may receive inadequate amounts. The high requirement of poultry and the low levels of Mn in many of the ingredients of poultry diets make Mn supplementation highly important (Hays and Swenson, 1985). Sources include whole grains, tea, legumes, nuts and seeds.

## Molybdenum (Mo)

Molybdenum is a component of several metalloenzyme including xanthine oxidase, aldehyde oxidase, nitrate reductase, and hydrogenase. Xanthine oxidase and aldehyde oxidase play a role in iron utilization as well as in cellular metabolism in electron transport. Xanthine oxidase is actively involved in the uptake and release of iron from ferritin in the intestinal mucosa and in the release of iron from ferritin in the liver, placenta, and erythropoietic tissues to the ferrous form (Hays and Swenson, 1985; Murray et al., 2000). It is readily absorbed from foods. Molybdenum is an important micronutrient both in animal and plant nutrition (Deosthale, 1980). In plants, it plays a role in nitrogen fixation and nitrate assimilation through nitrate reductase which is a key enzyme in the metabolic process in leguminous plants.

Dietary molybdenum affects copper metabolism in man (Deosthale and Gopalan, 1974). High intake of molybdenum can apparently precipitate copper deficiency in cattle and sheep (Underwood, 1971). The amount in the body is regulated by excretion in the urine and bile. Molybdenum is a cofactor for enzymes necessary for the metabolism of sulphur-containing amino acid and nitrogen-containing compounds present in DNA and RNA, the production of uric acid, and the oxidation and detoxification of various other compounds. Sardesai, (1993) reported that low molybdenum intake is a predisposing cause of renal xanthine calculi. Deficiency disease or symptoms are secondary to parenteral nutrition. It also causes gout. Sources include whole grains, dairy products, organ meats and legumes.

## *Selenium (Se)*

Selenium is a constituent of glutathione peroxidase (Murray et al., 2000). It is a constituent element of the entire defence system that protects the living organism from the harmful action of free radicals. Organic selenium is more thoroughly resorbed and more efficiently metabolized than its inorganic equivalent, which is poorly resorbed and acts more as a prooxidant provoking glutathione oxidation and oxidative damage to the DNA (Levander, 1983; Schrauzer, 2000; Wycherly et al., 2004). Se is a synergistic antioxidant with vitamin E. Its activity appears to be closely related to the antioxidative properties of _-tocopherol (vitamin E) and coenzyme Q (ubiquinone). It enhances the overall activity of the _- ketoglutarate oxidase system, probably by affecting the decarboxylation reaction. Organic and inorganic selenium compounds function in preventing certain disease conditions that have in the past been associated with vitamin E deficiency. Se prevents liver necrosis in rats, white muscle disease in lambs, and exhudative diathesis in chicks. Se protects the organism from oxidative damage to cell membranes by destroying $H_2O_2$, whereas vitamin E protects against damage by preventing the formation of the lipid hydroperoxides (Hays and Swenson, 1985).

The most important metabolic role of selenium is manifested in the activities of the selenoenzymes glutathione peroxidase (GSH-Px) and thioredoxin reductase. According to the reports by Surai, (2000), over the first few days after hatching, chicks mainly depend on selenium supplies stored in the liver. The activity of selenium and an adequate supply of vitamin E and some enzymes like GSH-Px, superoxide dismutase (SOD), catalase (CAT) protect chicks from numerous diseases like encephalomalacia, exudative diathesis and muscular dystrophy (Combs Jr. 1981; Hassan et al., 1990). The activity of SOD in the cells and in the extracellular fluid is very important in the prevention of diseases closely associated with oxidative stress, for example, cardiovascular diseases, Alzheimer's disease, Parkinson's disease and many other diseases (Pollack and Leeuwenburgh, 1999).

Se deficiency results in white muscle disease, an illness that cause high mortality in young calves and lambs. The clinical signs are myopathy that affects both the heart and skeletal muscle, frequently accompanied by abnormal calcification. White muscle disease is not a serious problem in areas where the soil is high in selenium. In the mild acute form, Se deficiency interferes with the normal growth processes of sheep and cattle. Se deficiency also disrupts the normal reproductive process, apparently affecting ovulation and fertilization, resulting in a higher incidence of embryonic mortality. Se deficiency has also been associated with a high incidence of retained placenta, resulting in delayed onset of oestrus and impaired conception (Hays and Swenson, 1985). Se deficiency in growing chickens cause exudative diathesis, signs like unthriftiness, ruffled feathers appear early at the growing stage. Egg production and feed conversion are adversely affected (Merck, 1986). In plants, Se is incorporated primarily by replacement of sulfur in the amino acids cystine and methionine. Toxic levels in plants result in blind staggers in horses and in the sloughing of hair and hoofs in horses and cattle. The animals become lame, and death in such cases is mainly caused by starvation resulting from movement impedement. In pigs, Se deficiency results in liver necrosis, hepatosis dietetica, resulting in high mortality. Mulberry heart disease in swine has been attributed to Se deficiency and is complicated by other factors other than Se deficiency (Hays and Swenson, 1985). In humans, toxic levels in some soils and megadose supplementation induces hair loss, dermatitis, and irritability (Murray et al., 2000). Deficiency disease or symptoms are also secondary to parenteral nutrition, protein-energy malnutrition. There is a considerable variation in the availability of selenium in different feedstuffs. Even in the presence of adequate levels of vitamin E, poultry rations still require an adequate level of selenium in the feed but care should be taken to prevent selenium toxicity. Sources include fishmeal, seafoods, dried brewer's yeast, kidney, liver, eggs and grains.

## *Silicon (Si)*

Silicon is one of the most abundant elements in plant and animal tissue. Silicon was confirmed as an essential element in 1972, when rats and chickens fed on a silicondepleted diet showed reduced weight gains and pathological changes in the formation and structures of Soetan et al. 213 collagenous connective tissues and bone (Birchall et al., 1996). Si is an essential component of certain mucopolysaccharides, hyaluronic acid and chondroitin-4-sulfate, which are important constituents of connective tissue. It appears to function as a biological cross-linking agent contributing to the structure and resiliency of connective tissue (Carlisle, 1974; Hays and Swenson, 1985). It plays a role in calcification of bone and in glycosaminoglycan metabolism in cartilage and connective tissue (Murray et al., 2000). Studies have shown that silicic acid, $Si(OH)_4$, the form in which silicon exists in physiology, interacts with aqueous aluminium (Al) species to form hydroxyl aluminosilicates that can have low bio-availability and toxicity and it is also established that the gastrointestinal absorption of aluminium is greatly reduced in the presence of silicic acid and also that the intake of dietary silicic acid also influences the excretion of aluminium via the kidneys so that Si appears to be profoundly involved in aluminium homesostasis (Birchall et al., 1996).

There are also strong indications that silicic acid promotes copper utilization and that the observed effects of silicon deficiency on collagen and osteogenesis may be due to low copper utilization (Birchall et al., 1996). It is however uncertain whether reduced copper utilization is caused by aluminium so that it is enhanced by silicic acid via the latters interaction with aluminium. Silicon and its salt are poorly soluble in water and only trace amounts are present in tissues. Herbivorous animals consume relatively large quantities of silica (silicon dioxide) daily, and most of the insoluble silica passes unabsorbed through the alimentary tract, but appreciable amounts are absorbed and excreted in the urine. Although, the silica is eliminated without side-effects, in some animals a portion is deposited as granules in the bladder, kidney, or urethra to form uroliths or calculi and these can block urine passages, which can result in death. These calculi may be composed of various minerals, especially Mg, P and Si. Silicon is essential for the normal growth and skeletal development in rats and chicks (Schwarz, 1974). Deficiency disease or symptoms implies impairment of normal growth. Silicosis due to long-term inhalation of silica dust arises from silicon toxicity. Sources include whole grain products, root vegetables.

## *Zinc (Zn)*

Zinc is distributed widely in plant and animal tissues and occurs in all living cells. It functions as a cofactor and is a constituent of many enzymes like lactate dehydrogenase, alcohol dehydrogenase, glutamic dehydrogenase, alkaline phosphatase, carbonic anhydrase, carboxypeptidase, superoxide dismutase, retinene reductase, DNA and RNA polymerase. Zn dependent enzymes are involved in macronutrient metabolism and cell replication (Hays and Swenson, 1985; Arinola, 2008). Carbonic anhydrase is present in erythrocytes, kidney tubules, gastrointestinal mucosa and glandular epithelium. Theprimary roles of zinc appear to be in cell replication and gene expression and in nucleic acid and amino acid metabolism. Vitamins A and E metabolism and bioavailability are dependent on zinc status (Szabo et al., 1999). It is necessary for fertility of mice. It is also required for normal testicular development (Merck, 1986) and for functions of the taste buds. It is needed for tissue repair and wound healing, plays a vital role in protein synthesis and digestion, and is necessary for optimum insulin action as zinc is an integral constituent of insulin. It is an important constituent of plasma (Malhotra, 1998; Murray et al., 2000).

In birds, zinc is required primarily for the growth and development of the skeleton, the formation and maintenance of epithelial tissue and for egg production (Gordon, 1977). Elemental zinc prevents and cures parakeratosis (thickening or hyperkeratinization of the epithelial cells of the skin and oesophagus) in swine and it prevents a similar disease in chicks. Excess calcium in diet however hastens the onset of parakeratosis. Formation of zinc fingers in nuclear receptors for steroid-thyroid, calcitriol receptors, gene expression, essential in protein synthesis, involves in the storage and release of insulin, growth and repair of tissues, development of sex organs, needed in the enzymes required for the synthesis of DNA and RNA, mobilization of vitamin A from the liver and stabilization of cell membranes. It is present in meat and other protein foodstuffs, but intestinal absorption is affected by other dietary constituents. Absorbed zinc enters the liver where it is incorporated into zinc metalloenzymes and exported to peripheral tissue in plasma, bound to albumin. A large percentage, about 90%, of the total plasma zinc concentration is associated with albumin, $<10\%$ with alpha-2 macroglobulin, and a small amount, $<1\%$, complexed to amino acids and other low molecular weight species. A high dietary iron intake can decrease zinc absorption.

Zinc homeostasis is achieved by regulation of enterohepatic re-circulation. An amount of zinc equivalent to the total absorbed zinc is re-excreted into the gut in intestinal fluids. In normal health zinc output by the gut is equal to the total dietary intake. Urinary excretion of zinc is low and does not vary markedly with dietary supply. It is increased in catabolic states, by certain drugs and/or chelating agents. Galvanized cages and troughs provide poultry with appreciable amounts of zinc and to a certain extent, this have been responsible for the infrequent occurrence of zinc deficiency in the field (Gordon, 1977). However, with the increasing use of plastics instead of galvanized steel, it is now important to ensure that poultry diets contain adequate amounts of this element and some are now supplemented with zinc oxide or carbonate. In humans, deficiency disease or symptoms include hypogonadism, growth failure, impaired wound healing, decreased taste and smell acuity, secondary to acrodermatitis enteropathica, parenteral nutrition (Murray et al., 2000). In poultry, Zn deficiency causes growth retardation and poor feather development. The hock joints may become enlarged and the long bones shortened and thickened. The skin on the footpads may become dry and thickened with fissures and hyperkeratosis develops (Merck, 1986). In mature hens, Zn deficiency reduces egg production and hatchability. Embryos show a wide range of skeletal abnormalities, including micromelia, curvature of the spine and shortened, fused thoracic and lumbar vertebrae (Merck, 1986).

Zinc deficiency in pigs causes a marked depression of appetite, growth rate and parakeratosis. In the young birds a deficiency of zinc is characterized by poor growth, severe dermatitis, especially of the feet and poor feathering, abnormal respiration, skeletal abnormalities causing leg weakness and ataxia. The long bones are shortened and thickened and are sometimes crooked and the joints are enlarged and rigid (Gordon, 1977). A necrotic dermatitis appears, particularly on the legs and feet, and feather development is impaired by hyperkeratinization of the epidermis. In severely deficient embryos, there are gross faults in the development of the skeleton and entire limbs may be absent. The biochemical lesions causing this syndrome in the osteogenic processes are probably due partly from a reduction in the activity of alkaline phosphatase which is a zinc enzyme (Gordon, 1977). Toxicity disease or symptoms of zinc in humans include gastrointestinal irritation, vomiting, decreased immune function and a reduction in high density lipoprotein (HDL) cholesterol. Higher dietary levels of Zn are required in the presence of phytic acid to prevent parakeratosis and allow for normal growth (Sidhu et al., 2004).

The oxide, carbonate, and sulfate forms of zinc are efficiently utilized, whereas the sulfide form is poorly utilized. Zn, either ingested or injected, is primarily excreted in the faeces. The Zn found in the faeces consist mainly of unabsorbed dietary Zn, and the balance is from pancreatic excretions. Urinary excretion of Zn and several other metals are increased if chelating agents such as ethylenediaminetetraacetic acid (EDTA) are administered in combination with zinc (Hays and Swenson, 1985). Sources include red meat, fish meals, liver, eggs, dairy products, vegetables and some sea foods.

## Fluorine (F)

Fluorine is widely but unevenly distributed in nature. It increases hardness of bones and gives strength to the enamel tissues of the teeth. It is also essential for rat growth. It plays a role in the prevention and treatment of dental caries. Fluoride in saliva reduces cavities by reducing acid produced by bacteria and by increasing enamel remineralization after acid exposure. Fluoride has great affinity for calcium and so it is associated with calcification of bones and teeth (Malhotra, 1998; Murray et al., 2000). Fluoridation is the addition of fluoride saltsto water. Low fluoride intake had been reported to cause growth retardation, reduces fertility, and results in anaemia (Messer et al., 1974; Cerklewski, 1998). Fluoride is incorporated into the tooth enamel crystals, where it forms the compound fluorhydroxyapatite. This compound is more resistant to acidic materials than the hydroxyapatite, which used to cover the tooth before. Deficiencies of fluorine seem to increase the incidence of dental caries. It is also involved in osteoporosis.

Excessive intake of fluorine gives rise to fluorosis. Excessive concentrations of fluorine in drinking water causes corrosion of the enamel of the teeth, a process called mottling. Fluorine is an enzyme inhibitor and when this happens in odontoblasts and osteoblasts, teeth and bone deformities occur. Exostoses of the jaw and long bones develop and the joints become thickened and ankylosed, resulting in lameness. This could further result in reduction in food consumption, accompanied by reduced growth or weight loss (Hays and Swenson, 1985). No single symptom of fluorosis is a proof of toxicity. The chemical analysis of bone, teeth, and urine, with evidence of chalkiness, mottling, erosion of the enamel, enamel hypoplasia, and excessive wear of the teeth, as well of evidence of anorexia, inanition, and bone changes, are all necessary in diagnosing fluorine toxicosis. Sources include fluorinated drinking water, tea, toothpaste, marine fish consumed with bones, food cooked with Teflon utensils can pick up fluoride, leading to increase in content.

## Sulfur (S)

Sulfur is present in three amino acids which are cystine, cysteine and methionine. Connective tissue, skin, hair and nails are rich in sulfur. Also, thiamine and biotin (member of vitamin B complex) and coenzyme A contain sulfur in their molecules (Malhotra, 1998). Proteins vary widely in sulfur content, depending on their amino acid composition. The preformed amino acid and sulfur deficiencies are reflected as sulfur-containing amino acid deficiencies. Ruminants, that depend largely on nonprotein nitrogen sources, such as urea, biuret, or ammonium phosphate, may need supplemental inorganic sulfur, which is utilized by the microorganisms for synthesis of methionine and cystine. Inadequate sulfur intake reduces the feed intake, digestibility, rate of weight gain, and milk production (Hays and Swenson, 1985).

Taurine and sulfuric acid are the end products of sulfur metablosim. The sulfuric acid is either neutralized and excreted as inorganic sulfates in the urine or conjugated with phenol, glucuronic acid, or indoxyl. The taurine is conjugated with cholic acid and excreted in the bile. The other sulfur compounds of biological significance are cyanate (SCN) in saliva and other fluids, ergothioneine of the red blood cells, glutathione, present in all cells, and chondroitin sulfate, which serves a structural function in Soetan et al. 215 cartilage, bone, tendons, blood vessel walls etc. The primary function of sulphur is as the disulfide linkage, -SS-, of such organic compounds (Hays and Swenson, 1985). Dietary sources of sulfur are diets which are adequate in protein and these meet the daily requirement of sulfur.

### Iodine (I)

Iodine is a trace element that is essential for the synthesis of thyroid hormones in vertebrates, although iodoproteins are are present in invertebrates (Frieden, 1984). Some regions of the world are naturally deficient in iodine due to the low availability of iodine from their soil or other climatic and environmental factors affecting iodine availability.

Several goiter, cretinism and the other iodine deficiency disorders (IDD) have their main cause in the lack of availability of iodine from the soil which is linked to a severe limitation of food exchanges (Thilly et al., 1992). Severe selenium deficiency have also been associated with IDD in the human population, whereas in animals, it has been proven to play a role in thyroid function either through a thyroidal or extrathyroidal mechanism. The former involves oxidative damages mediated by free radicals, whereas the latter implies an inhibition of the deiodinase responsible for the utilization of T4 and T3 and is then concluded that: (i). goiter has a multifactorial origin (ii). IDD are an important public health problem and (iii). IDD are a good model to study the effects of other trace elements whose actions in many human metabolisms have been somewhat underestimated. Not less than 5% of the world's population have goiters and many of these are associated with other disorders and constitute a major public health problem (Gaitan, 1990).

## The Biochemical Functions of the Other Mineral Elements

### Tin (Sn)

Tin is required for normal growth in the rat and there is no specific clinical or pathological changes in the animal (Schwarz, 1974). It also improved the pigmentation of incisors in the rats.

### Bromine (Br)

Bromine increases the growth rate of chicks and mice (Huff et al., 1956). It has been reported that thyroid glands of rabbits concentrate bromine during periods of dietary iodine deficiency. This then suggests that the thyroid gland does not distinguish perfectly between iodine and bromine. However, bromine does not preventthe occurrence of goiter, and it is rapidly replaced by iodine when the latter is restored to the diet. According to Vobecky et al., (1996), in experiments with rats, it has been found that at enhanced intake of bromide, bromine does not replace chlorine in the thyroid, it replaces iodine. They further reported that more than one third of the iodine content in the thyroid was replaced by bromine.

## *Strontium (Sr)*

The omission of strontium caused an impairment of the calcification of the bones and teeth and a higher incidence of carious teeth. 90Sr is one of the most abundant and potentially hazardous radioactive byproducts of nuclear fission and plants are more efficient than animals in the absorption of strontium. Radioactive strontium is absorbed and deposited in tissues especially the bones, and is also readily transmitted to the foetus and secreted in the milk (Hays and Swenson, 1985). Stronium is preferentially excreted, especially in the urine, thereby providing some means of protection against 90Sr.

## *Nickel (Ni)*

Nickel is an essential element in animals (Nielsen, 1976). It has been speculated that nickel may play a role in the maintenance of membrane structure, control of prolactin, nucleic acid metabolism or as a cofactor in enzymes. It appears that most dietary intakes would provide sufficient amounts of this element (Hurley, 1976). Nickel (Ni) is an in nickel that is, containing $<0.05$ ppm developed increased skin pigmentation of the legs, swollen hocks and thickening of the legs near the joints. These signs were not apparent in chicks fed the same diet supplemented with 3-5 ppm nickel (Nielsen, 1974). In a trace element-controlled environment, chicks unsupplemented with nickel have shorter, thicker legs, lower haematocrit and plasma cholesterol levels and and higher liver cholesterol levelsthan control chicks. In rats, there is slower growth rate and post-natal mortality. In swine, there is impaired reproduction, abnormal haircoats, and poor growth of offspring (Anke et al., 1974; Tejada-Jimenez et al., 2009).

## Arsenic (As)

Arsenic stimulates the growth of tissue cultures (Nandi et al., 2006) and the beneficial effects of various organic arsenicals on the health and performance of pigs and poultry had been reported. Its action is similar to that of antibiotics and appears to be mainly related to the control of harmful intestinal micro-organisms (Hays and Swenson, 1985). The fact that cadmium in specific association with protein macromolecules had been fractionated from the horse kidney cortex suggests a biological role for cadmium (Kagi and Vallee, 1960).

## Rubidium (Rb)

Rubidium is found in animal tissue and it resembles potassium in its distribution and excretory pattern. Relatively high levels can be found in the soft tissue, while the skeletal tissue contains low level. The additions of rubidium or cesium (Cs) to potassium-deficient diets prevent the lesions characteristic of potassium depletion in rats (Follis, 1943) and supports near normal growth for short periods of time (Heppel and Schmidt, 1938).

## Vanadium (V)

Vanadium compounds have been demonstrated to have effects on glucose metabolism *in vitro* (Frieden, 1984) and also in lipid metabolism (Hopkins and Mohr, 1974). Vanadium also has a wide range of other biological activities like effects on ribonuclease, alkaline phosphatase, adenyl cyclase, NADH oxidase and phosphofructokinase (Ramasarma and Crane, 1981). Vanadium has been shown to be nutritionally essential for both the chick and the rat (Hopkins and Mohr, 1971; Strasia, 1971; Nielsen and Ollerich, 1973; Hopkins and Mohr, 1974). Hopkins and Mohr, (1971) reported reduced growth of wing and tail feathers in chicks fed a diet containing less than 10mg vanadium per gram. Bone development was also abnormal. In rats, reduced growth and increased packed cell volume as well as increased iron in blood and bone were observed (Strasia, 1971). Reproductive performance was also impaired in rats. However, rats fed purified diets low in vanadium responded to dietary vanadium supplementation with increased weight gain (Strasia, 1971) and with improved reproductive performance (Hopkins and

Mohr, 1974). A precise biochemical or physiological role for vanadium (V) is yet to be determined, but it has long been suspected because of its effects on phospholipids oxidation and inhibition of cholesterol synthesis. It is necessary for normal feather development in chicks and for normal growth in chicks and rats and the required level appears to be very low, $< 0.1$ to 2 ppm (Hays and Swenson, 1985). Essentiality of vanadium in man has been hypothesized but not demonstrated (Hopkins and Mohr, 1974).

## Cadmium (Cd)

The possible effects of the general population of long term, low level exposure to cadmium have been of concern recently (Asagba, 2009), because cadmium isreadily distributed in tissues after exposure and it inhibits antioxidant enzymes (Gupta et al., 1991; Bagchi et al., 1996; Asagba and Eriyamremu, 2007; Chater et al., 2009) and this inhibition can lead to increased oxidative stress which may result in membrane damage and loss of membrane-bound enzymes like ATPases (Figueiredo- Pereira et al., 1998; Asagba et al., 2004; Asagba and Obi, 2005; Galazyn-Sidorczuk et al., 2009). Liver and kidney are considered as the main target organs in acute and chronic Cd exposure (Sarker et al., 1995; Asagba and Obi, 2000). Other tissues involved in Cd toxicity are the testis, heart, bone, eye and brain (WHO, 1992; Pal et al., 1993; Lall et al., 1997; Asagba et al., 2002; Eryamremu et al., 2005; Asagba 2007). The role of diet in absorption and toxicity of oral cadmium have been reviewed by (Asagba, 2009).

## Lead (Pb)

Lead is an ubiquitous environmental and industrial pollutant that has been detected in every facet of environmental and biological systems. Lead can be found in water pipes, insecticides, lining of equipment where corrosion resistance and pliability are required, in petroleum refining, in construction, gun bullets, x-ray and atomic radiation protection (Ait et al., 2009) and the manipulation of lead for these uses had caused lead contamination of air, dust and soil (Kiran et al., 2008). Reports suggest that there is a close relationship between cases of declining reproductive health and environmental pollutants like lead (Bonde et al., 2002), while human exposure to lead continues to be a serious public health problem (Pirke et al., 1998). According to Satoskar et al. (1999), chronic lead poisoning is commonly seen in young children arising from sucking lead paint or lead toys and is also seen in workers engaged in printing, paint and petroleum industries.

Reproductive dysfunction by lead has distinct morphological and biochemical features such as disorganized epithelia, decreased sperm quality and altered sperm morphology and low androgen levels (Hsu et al., 1997; Alexander et al., 1996). In the animal model, lead has a primary toxic effect on the hypothalamic-pituitary unit, a primary effect on the testes and acts at all levels of the reproductive axis (Sokol et al., 2002). Male rats exposed to lead acetate have been reported to show a significant decrease in the weight of both the testes, epididymis (Wang et al., 2008; Smith et al., 2008) and pituitary (Cooper et al., 1986; Coffigny et al., 1994) and this weight reduction in sex glands and pituitary was accompanied by an alteration of the normal histological structure. There are reports that some metal such as lead, cadmium, arsenic and mercury can affect male reproductive functions including sperm counts (Chia et al., 1992), motility and morphology (Meeker et al., 2008) and spermatogenesis (Telisman et al., 2007).

The coastal zone receives a large amount of metal Soetan et al. 217 pollution from agricultural and industrial activities (Usero et al., 2005). Pollution by heavy metals is a serious problem due to their toxicity and ability to accumulate in the biodata (Islam and Tanak, 2004). There is a general concern about the impact of metals in the aquatic environment (Grosell and Brix, 2005). Lead (Pb), Cadmium (Cd), copper (Cu), zinc (Zn) are the most present trace heavy metals in the wastewater of the gulf (Semadi and Deruelle, 1993; Abdennour et al., 2000; 2004). Zn and Cu are used in small amounts as fertilizers in some soils deficient in these elements, while Zn, Cu and Cd are constituents of some fungicides. In addition, Cd and Zn occur as contaminants of phosphoric fertilizers (Migon et al., 1991).

According to Beldi et al. (2006), biomonitoring programs based on measures of contaminants in marine organisms are interesting from a human health point of view. Many factors such as season and physiological state or size of animals could affect the bioaccumulation of heavy metals in molluscs (Romeo and Gnassia-Barelli, 19988). It is generally agreed that heavy metal uptake occurs mainly from water, food and sediment (Beldi et al., 2006). However, effectiveness of metal uptake from these sources may differ in relation to ecological needs and metabolism of animals and concentrations of the heavy metals in water, food and sediments as well as some other factors such as salinity, temperature and interacting agents (Roesijiadi and Robinson, 1994).

# THE BIOCHEMICAL FUNCTIONS AND THE DEFICIENCY SYMPTOMS OF THE MINERAL ELEMENTS IN PLANTS

The essential elements that are required for the healthy growth of higher plants are carbon, hydrogen, oxygen and those listed in Table 1 (Williams, 1992). Plants take in carbon in the form of atmospheric carbon dioxide, obtain hydrogen and oxygen from water and take up the remaining elements from the soil. Mineral nutrients are essential chemical elements absorbed from the soil in the form of inorganic ions. More than 50 chemical elements have been identified in plants among the inorganic substances present in plants, although it is unlikely that all these elements are essential. Roots can absorb minerals selectively, thus, enabling the plant to accumulate essential elements that may be present in the soil in very minute quantities. To a certain extent, however, the minerals in a plant reflect the composition of the soil in which the plant is growing. For example, plants growing on mine tailings may contain gold or silver.

Minerals play essential role in enzyme activity in plants. For instance, the effects of cobalt on enzyme activity in plants have been reported in many publications (Frieden, 1984). Cobalt additions increased the activity of catalase in cotton and of catalase, ascorbic oxidase, polyphenoloxidase and peroxidase in grapevines. Cobalt also enhanced the activity of catalase and dehydro-genase inpotatoes and of catalase and peroxidase in tomatoes. In legumes, cobalt was found to exert to exert a positive effect on the enzyme activity of nodules. Cobalt has also been reported to improve the drought-resistance of barley, corn, oats and soybeans, and also increases the rates of synthesis and migration of carbohydrates from the leaves to the fruit-bearing organs (Frieden, 1984). The fact that all soil-grown plants contain silicon and that large quantities may be present in such species as rice, other grasses and horsetails had led to the suggestion that silicon has beneficial effects on plant growth and crop yield. It has also been suggested in some circumstances to promote resistance to insect and fungal attack and this has been reviewed by (Lewin and Reimann, 1969).

Plant uptake of silicon is governed by its concentration in the soil solution and other factors like the nature of the plant (Frieden, 1984). Boron is also believed to be important mineral element for all vascular plants (Lewis, 1980). It is the only element known to be required for higher plants for which there is no role in humans and animals (Mondy and Munshi, 1993). Lee and Aronoff, (1967) have proposed that the primary function of boron is based on its capacity to form stable 6-P-gluconateborate complexes and hence restrict both the influx of substrate into the pentose-phosphate pathway and the synthesis of phenols resulting in increases in glycolysis and the synthesis of hemicelluloses and other related cell wall materials. In plants, boron is thought to regulate the metabolic pathways leading to phenol synthesis and is also thought to be involved in the pathway leading to the synthesis of lignin. Lewis (1980) proposed that boron's primary role in plants is in the biosynthesis of lignin and the differentiation of xylem tissue. Compounds like leucoanthocyanins, flavonols, flavonones and flavonol-3- glucosides, have been reported to accumulate in borondeficient plant tissue (Rajaratnam and Lowry, 1974; Shkol'nik and Abysheva, 1975). The interralationships among boron, lignification, peroxidase enzymes and auxin in plants have been discussed by (Lewis, 1980). In addition, he discussed the similarities by which wounding a vascular plant, microbial infection of the plant and boron deficiency bring about changes in the level of phenolic compounds and in indole acetic acid (IAA) metabolism (Lewis, 1980).

It has been proposed that a study of the evolution of vascular plants may provide a possible key to the role of boron in plant metabolism. Lewis (1980a) had proposed that the development of an essential role for boron was a prerequisite to the evolution of vascular plants. Augsten and Eichhorn (1976) demonstrated that a suboptimal level of boron caused an increase in the photosynthetic rate of *Wolffia arrhiza*. Leaves usually exhibit the first symptoms of boron toxicity which include yellowing of the leaf tip, with the chlorosis subsequently progressing along the leaf margin and then spreading into the blade.

Necrosis of the chlorotic tissue follows, followed by leaf abscission (Wilcox, 1960). As a result of this, boron toxicity results in a loss of plant toxicity. The most common symptoms of mineral deficiency are stunted growth and discoloured leaves. The symptoms of a mineral deficiency depend partly on the function of that nutrient in the plant. For example, a deficiency of magnesium, an ingredient of chlorophyll, causes yellowing of the leaves or chlorosis and in some cases, the relationship between a mineral deficiency and its symptoms is less distinct. For instance, iron deficiency can cause chlorosis even though chlorophyll contains no iron, because this metal is required as a cofactor in one of the steps of chlorophyll synthesis (Roberts, 1985).

Mineral deficiency symptoms depend not only on the role of the nutrient in the plant but also on its mobility within the plant. If a nutrient moves about freely from one part of the plant to another, symptoms of a deficiency will show up first in older organs. This is because young, growing tissues have more ability/power than old tissues to draw the nutrients that are in short supply. In contrast, the deficiency of a nutrient that is relatively immobile within the plant will first affect young parts of the plant because older tissues may have adequate amounts of the mineral, which they are able to retain during periods of short supply. The symptoms of a mineral deficiency are often distinctive enough for a plant physiologist or farmer to diagnose its cause. Deficiencies of nitrogen, potassium and phosphorus are the most common problems, while shortages of micronutrients are less common and tend to be geographically localized because of deficiencies in soil composition. However, overdoses of some micronutrients can be toxic to plants (Roberts, 1985). Table 1 shows the macronutrients and micronutrients in plants, the form available to plants, their major functions and deficiency symptoms.

# CONCLUSION

In view of the importance of these macro and micro mineral elements in the nutrition of humans, animals and plants and their metabolic inter-relationships which influences other vital factors needed for the survival of living organisms like enzymes, anti-oxidants, vitamins etc, it is important to regularly obtain up-to-date information on the minerals content of water and the variously commonly consumed plant foods used for human and animal foods and feeds respectively e.g. legumes, cereals, fruits and vegetables. This is because of the effects of chemicals like pesticides, herbicides, fertilizers on the mineral contents of the soils where these plants are cultivated. Genetic, location and environmental factors could also influence the levels of the mineral elements in plants.

# RECOMMENDATION

With the advancements in the field of analytical sciences, involving the use of more sophisticated equipment for the quantitative and qualitative analysis of foods/feeds. There is need to re-visit the studies/ investigations /assessments /evaluations, so as to re-validate the data on the mineral elements composition of human and animal diets, especially in the developing countries. The earlier information on these mineral elements were based on analysis employing less sensitive methods which may not be reliable. Further research is required to find out the interactions between the mineral elements, for example, the interaction between silicon, aluminium and copper. There is also the need to evaluate the effects of different processing methods on the minerals content of water and plant foods. This will go a long way in reducing the problem of malnutrition caused by inadequate food supply, consumption of unbalanced diets and inadequate nutrition education.

# REFERENCES

1. Abdennour C, Khelili K, Boulakoud MS, Rainbow PS (2004). Trace metals in marine, brackish and fresh water prawns (Crustacea, Decapoda) from NorthEast Algeria. Hydrobiologia 432: 217-227.
2. Abdennour C, Smith BD, Boulakoud MS, Samraoui B, Rainbow PS (2000). Trace metals in shrimps and sediments from Algerian water. J. Catalog. Mat. Env. 3: 9-12.
3. Albion Research Notes (1996). A compilation of vital research updates on human nutrition, 5: 2, Albion Laboratories, Inc. May, 1996.

4. Aletor VA.Omodara OA (1994). Studies on some leguminous browse plants, with particular reference to their proximate, mineral and some endogenous anti-nutritional constituents. Anim. Feed Sci. Technol. 46: 343-348.

5. Alexander BH, Chechoway H, Van Netten C, Muller CH, Ewers TG, Kaufman JD (1996). Semen quality of men employed at a lead smelter. Occup. Environ. Med. 53(6): 411-416.

6. Anke M, Grun M, Dittrick G, Groppel B, Hennig A (1974). In: Trace Element Metabolism in Animals-2, WG Hoekstra, JW Shuttle, HE Ganther and W Mertz Editors. University Park Press, Baltimore pp. 715-718.

7. Arinola OG (2008). Essential trace elements and metal binding proteins in Nigerian consumers of alcoholic beverages. Pak. J. Nutr. 7(6): 763-765.

8. Arinola OG, Badmos SO, Ademowo OG (2008b). Trace elements and anti-oxidant status in Gravid BALB/c mice infected with *Plasmodium yoelii* malaria parasites at different gestational periods. Pak. J. Nutr. 7(6): 757-762.

9. Arinola OG, Nwozo SO, Ajiboye JA, Oniye AH (2008c). Evaluation of trace elements and total antioxidant status in Nigerian cassava processors Pak. J. Nutr. 7(6): 770-772.

10. Arinola OG, Olaniyi JA, Akiibinu MO (2008a). Evaluation of antioxidant levels and trace element status in Nigerian sickle cell disease patients with *Plasmodium* parasitaemia. Pak. J. Nutr. 7(6): 766-769.

11. Asagba SO (2007). Alterations in activities of tissue enzymes in oral cadmium toxicity. Niger. J. Sci. Env. 6: 91-102.

12. Asagba SO (2009). Role of diet in absorption and toxicity of oral cadmium. Afr. J. Biotechnol. 8(25): 7428-7436.

13. Asagba SO, Eriyamremu GE (2007). Oral cadmium exposure alters haematological and liver function parameters of rats fed a Nigerianlike diet. J. Nutr. Envt. Med. 16(1-3): 267-274.

14. Asagba SO, Eriyamremu GE, Adaikpoh MA, Ezeoma A (2004). Levels of lipid peroxidation, superoxide dismutase and Na+/K+-ATPase in some tissues of rats exposed to a Nigerian diet and cadmium. Biol. Trace Elem. Res. 100(1): 075-086.

15. Asagba SO, Isamah GK, Ossai EK, Ekakitie AO (2002). Effect of oral exposure to cadmium on the levels of vitamin A and lipid peroxidation in the eye. Bull. Environ. Contam. Toxicol. 68: 18-21. Soetan et al. 219

16. Asagba SO, Obi FO (2005). A comparative evaluation of the biological effects of environmental cadmium contaminated control diet and laboratory cadmium supplemented test diet. Biometals 18: 155-161.

17. Asagba SO, Obi FO (2000). Effect of cadmium on kidney and liver cell membrane integrity and antioxidant enzyme status: implications for Warri River cadmium level. Trop. J. Environ. Sci. Health 3(1): 33-39.

18. Augsten H, Eichhorn B (1976). Biochemistry and physiology of the effect of boron in plants, Biol. Rundsch. 14: 268-285.

19. Bagchi D, Bagchi M, Hassoun EA, Stohs SJ (1996). Cadmium-induced excretion of urinary lipid metabolites, DNA damage, glutathione depletion and hepatic lipid peroxidation in Sprague-Dawley rats. Biol. Trace Elements Res. 52(2): 143-154.

20. Basargin NN, Peregudora LA (1969). Zinc and molybdenum content of different varieties of spring and winter wheat. Vop. Pttan, 28: 65.

21. Baser BL, Deo R (1967). J. Ind. Soc. Soil Sci. 15: 245. as cited by Kanwar JS and Randhawa NS (1974). Micronutrient Research in Soils and Plants in India (A Review). 2nd Edn, pp. 41-78. Indian Council of Agricultural Research, New Delhi.

22. Batra J, Seth PK (2002). Effect of iron deficiency on developing rat brain. Indian J. Clin. Biochem. 17(2): 108-114.

23. Beard JL (1999). Iron deficiency and neural development: an update- Arch. Latinoam Nutr. 49(3 Suppl. 2): 34-39.

24. Beard JL (2001). Iron biology in immune function, muscle metabolism and neuronal functioning, J. Nutr. 131: 5685-5695.

25. Belakbir A, Ruiz JM, Romero L (1998). Yield and fruit quality of pepper (*Capsicum annum* L.) in response to bioregulators. Hort. Sci. 33: 85- 87.

26. Beldi H, Gimbert F, Maas S, Scheifler R, Soltani N (2006). Seasonal variations of Cd, Cu, Pb and Zn in the edible Mollusc

*Donax trunculus* (Mollusca, Bivalvia) from the gulf of Annaba, Algeria. Afr. J. Agric. Res. 1(4): pp. 085-090.

27. Bendich A (1990). Antioxidant micronutrients and immune responses. Ann. N.Y. Acad. Sci. 587: 180.
28. Birchall JD, Bellia JP, Roberts NB (1996). On the mechanisms underlying the essentiality of silicon-interactions with aluminium and copper. Coordination Chemistry Reviews (Abstract), 149: 231-240.
29. Blood DC, Radostits OM (1989). Veterinary Medicine, 7th Ed. Balliere Tindall, London pp. 589-630.
30. Bonde JP, Joffe M, Apostoli P, Dale A, Kiss P, Spano M (2002). Sperm count and chromatin structure in men exposed to inorganic lead: lowest adverse effect levels. Occup. Environ. Med. 59(4): 234-242.
31. Brown M (2003). Harnessing chromium in the fight against diabetes. Drug Discovery today 8: 962-963.
32. Burk RF, Foster KA, Greenfield PM, Kiker KW (1974). Binding of Simultaneously Administered Inorganic Selenium and Mercury to a Rat Plasma Protein. Proc. Soc. Exp. Biol. Med. 145: 782-785.
33. Carlisle EM (1974). Essentiality and function of silicon. In Hoekstra et al., eds. Trace Elements Metabolism in Animals, 2nd Ed. University Park Press, Baltimore. Cerklewski FL (1998). Fluoride: Essential or just beneficial. Nutr. Rev. 14: 475-476.
34. Chand M (1969). Effect of different levels of nitrogen application on the micronutrients removal by Mexican and local wheat varieties. M.Sc. Thesis. Punjab Agricultural University, Ludhiana.
35. Chandra RK (1990). Micro-nutrients and immune functions: An overview. Annal New York Acad. Sci. 587: 9-16.
36. Chater S, Douki T, Favier A, Sakly M, Abdelmelek H (2009). Changes in antioxidant status and biochemical parameters after oral cadmium administration in female rats. Acta. Biol. Hung. 60(1): 79-88.
37. Chakravarty I, Ghosh k (2000). Micronutrient malnutrition-present status and future remedy. J. Indian Med. Assoc. 98(9): 539-542.

38. Chia SE, Ong CN, Lee ST, Tsakok FH (1992). Blood concentrations of lead, cadmium, mercury, zinc and copper and human semen parameters. Arch. Androl. 29: 177-183.
39. Clauss M, Sauter-Loius C, Kienzle E (2007). Historical copper and manganese levels in cattle feeds in Bavaria, Germany. J. Nutr. Environ. Med. 16: 69-74.
40. Coffigny H, Thoreux-Manlay A, Pinon-Lataillade G, Masse R, Soufir JC (1994). Effects of lead poisoning of rats during pregnancy on the reproductive system and fertility of their offspring. Hum. Exp. Toxicol. 13(4): 241-246.
41. Combs Jr. GF (1981). Influence of dietary vitamin E and selenium onthe oxidant defence system of the chick. Poultry Sci. 60: 2098-2105.
42. Cooper RL, Goldman JM, Rehnberg GL (1986). Pituitary function following treatment with reproductive toxins. Environ. Health Persp. 70: 177-184.
43. Darby WJ (1976). Trace elements in human health and disease, Prasad AS.and Oberleas D. Eds (Academic Press, New York, San Francisco, London) 1: 17.
44. Deosthale YG (1990). Molybdenum content of some common Indian Foods. The Ind. J. Nutr. Dietet. 18: 15-19.
45. Deosthale YG, Belavady B (1978). Mineral and Trace Element Composition of Sorghum grain: Effect of Variety, Location and Application of the Nitrogen Fertilizer. Ind. J. Nutr. Dietet. 15: 302- 308.
46. Deosthale YG, Gopalan C (1974). The effect of molybdenum levels in sorghum (*Sorghum vulgare* Pers.) on uric acid and copper excretion in man. Brit. J. Nutr. 31: 451.
47. Eastmond DA, MacGregor JT, Slesinki RS (2008). Trivalent Chromium: Assessing the genotoxic risk of the essential trace element and widely used human and animal nutritional supplement. Crit. Rev. Toxicol. 38: 173-190.
48. Ellis DR, Salt DE (2003). Plants, Selenium and Human Health. Current Opinions in Plant Biology 6: 273-279.
49. El-Behairy F, El-Blassousy E, Diwany KM (1997). Manual of Paediatrics. University Book Centre p. 16. Eriyamremu GE, Asagba SO, Onyeneke EC, Adaikpoh MA (2005). Changes in

carboxypeptidase A, dipeptidase and Na+/K+-ATPase activities in the intestine of rats orally exposed to different doses of cadmium. Biometals 18: 1-6.

50. Eruvbetine D (2003). Canine Nutrition and Health. A paper presented at the seminar organized by Kensington Pharmaceuticals Nig. Ltd., Lagos on August 21, 2003.
51. Figueiredo-Pereira ME, Yakushin S, Cohen G (1998). Disruption of the intracellular sulhydryl homeostasis by cadmium-induced oxidative stress leads to protein thiolation and ubiquitination in neuronal cells. J. Biol. Chem. 273: 12703-12709.
52. Follis RH Jr. (1943). Histological effects in rats resulting from adding rubidium or cesium to a diet deficient in potassium. Am. J. Physiol. 138: 246-250.
53. Forbes RM, Parker HM (1977). Biological availability of zinc in and as influenced by whole-fat soy flour in rat diets. Nutr. Reports Int. 15: 681- 688.
54. Frieden E (1984). Biochemistry of the essential ultratrace elements. Plenum press, New York.
55. Gailer J, George GN, Pickering IJ, Madden S, Prince RC, Yu EY, Denton MB, Younis HS, Aposhian HV (2000). Structural basis of the antagonism between inorganic mercury and selenium in mammals. Chem. Res. Toxicol. 13(11): 1135-1142.
56. Gaitan E (1990). Goitrogens in food and water. Annual Rev. Nutr. 10: 21-37.
57. Galan P, Viteri F, Bertrais S, Czernichow S, Faure H (2005). Serum concentrations of beta carotene, vitamins C and E, zinc and selenium are influenced by sex, age, diet, smoking status, alcohol consumption and corpulence in a general French adult population. Eur. J. Clin. Nutr. 59: 1181-1190.
58. Galazyn-Sidorczuk M, Brzoska MM, Jurczuk M, Moniuszk-Jaconiuk J (2009). Oxidative damage to proteins and DNA in rats exposed to cadmium and or ethanol. Chem. Biol. Interac. 180(1): 31-38.
59. Gardea-Torresdey JL, Becker-Hapak MK, Hosea JM, Darnall DW (1990). Effect of chemical modification of algae carboxyl groups onmetal ion binding. Environ. Sci. Technol. 24: 1372-1378.

60. Gordon RF (1977). Poultry Diseases. The English Language Book Society and Bailliere Tindall, London.
61. Gorsuch TT (1959). Radiochemical investigations on the recovery for analysis of trace elements in organic and biological materials. Analyst 84: 135.
62. Grashuis J, Lehr JJ, Beuvery LLE, Beuvery-Asman A (1953). "De Schothorste" Inst. Moderne Veevoeding, Netherlands. Cited by EJ Underwood in Trace Elements in Human and Animal Nutrition, 2nd ed. Academic Press, New York, 1962, p. 197.
63. Grossell M, Brix KV (2005). Introduction to the special issue on mechanisms in metal toxicology. Aquat. Toxicol. 72: 3-4.
64. Gruden N (1977). Suppression of transduodenal manganese transport by milk diet supplemented with iron, Nutr. Metab. 21: 305-309.
65. Gupta S, Althar M, Behari JR, Srivastava RC (1991). Cadmiummediated induction of cellular defence mechanism: a novel example for the development of adaptive response against a toxicant. Ind. Health 29: 1-9.
66. Hassan S, Hakkarainen J, Jonson L, Tyopponen J (1990). Histopathological and biochemical changes associated with selenium and vitamin E deficiency in chicks. J. Vet. Med. Assoc. 37: 708-720.
67. Hays VW, Swenson MJ (1985). Minerals and Bones. In: Dukes' Physiology of Domestic Animals, Tenth Edition pp. 449-466.
68. Hegsted DM, Chichester CO, Darby WJ, McNutt KW, Stalvey RM, Stotz EH (1976). In: Present Knowledge in Nutrition (Nutrition Reviews'), Fourth Edition. The Nutrition Foundation, Inc. New York, Washington.
69. Heppel LA, Schmidt CLA (1938). Studies on the potassium metabolism of the rat during pregnancy, lactation and growth. Univ. Calif. (Berkeley) Pub. Physiol. 8: 189-205.
70. Hopkins LL Jr, Mohr HE (1971). The biological essentiality of vanadium, In: Newer Trace Elements in Nutrition. W Mertz, WE Cornatzer, Editors, Marcel Dekker, New York pp. 195-213.
71. Hopkins LL Jr, Mohr HE (1974). Vanadium as an essential nutrient, Fed. Proc. 33: 1773-1775.

72. Horsfall M Jnr, Abia AA, Spiff AI (2003). Removal of Cu(II) and Zn(II) ions from waste water by cassava (*Manihot esculenta* Cranz) waste biomass. Afr. J. Biotechnol. 2(10): 360-364.

73. Hsu PC, Liu MY, Hsu CC, Chen LY, Leon Guo Y (1997). Lead exposure causes generation of reactive oxygen species and functional impairment in rat sperm.Toxicology 122(12): 133-143.

74. Huff JW, Bosshardt DK, Miller OP, Barnes RH (1956). A nutritional requirement for bromine. Proc. Soc. Exp. Biol. Med. 92: 216-219.

75. Hurley LS (1976). Manganese and other Trace Elements. In: Present Knowledge in Nutrition (Nutrition Reviews'), Fourth Edition. The Nutrition Foundation, Inc. New York, Washington.

76. Igile GO (1995). Phytochemical and Biological Studies on some constituents of *Vernonia amygdalina* (compositae) leaves. Ph.D Thesis, Department of Biochemistry, University of Ibadan, Nigeria.

77. Islam MD, Tanaka M (2004). Impact of pollution on coastal and marine ecosystems including coastal and marine fisheries and approach for management: A review and synthesis. Mar. Pollut. Bull. 48: 624-649.

78. Juturu V, Komorowski JR (2003). Chromium Supplements, glucose and insulin responses. Am. J. Clin. Nutr. 78: 190.

79. Kagi JHR, Vallee BL (1960). Metallothionein: A cadmium- and zinccontaining protein from quine renal cortex. J. Biol. Chem. 235: 3460- 3465.

80. Kavanek M, Janicek G (1969). Effect of locality and variety on the content of some trace elements in oat grains. Shorn. Vys. Skoly Chemtechnol Praze E, 24: 65.

81. Kiran KB, Erika B, Rashidi M, Prabhakara RY, Sharada R, Rajanna B (2008). Lead-induced increase in antioxidant enzymes and lipid peroxidation products in developing rat brain, Biometals 21: 9-16.

82. Kirchgessner M, Schwarz FJ, Roth-Maier DA (1981). Changes in the metabolism (retention, absorption, excretion) of copper, zinc, and manganese in gravidity and lactation, in: Trace Element Metabolism in Man and Animals (TEMA-4), McC Howell J, Gawthorne JM and White CL (eds.), Australian Academy of Sciences, Canberra pp. 85-88.

83. Lall SB, Das N, Rama R, Peswin SS, Khattar S, Gulatti K, Seth SP (1997). Cadmium- induced nephrotoxicity in rats. Indian J. Exp. Biol. 35(2): 151-154.

84. Langard S (1980). Chromium, in Metals in the Environment, H.H.Waldron (ed,), Academic Press, London pp. 111-132.

85. Larkin EC, Rao CA (1990). Importance of fetal and neonatal iron; adequacy for normal development of central nervous system, In: Brain behavior and iron in the infant diet (Dobbing J ed), London, UK pp. 43-63.

86. Lease JG (1966). The efficiency of autoclaving sesame meat on its phytic acid content and on the availability of its zinc to the chicken. Poult. Sci. 45: 237 - 241.

87. Leach RM Jr. (1974). Biochemical role of manganese. In: WG Hoekstra *et al* eds., Trace Element Metabolism in Animals, 2nd ed. University Park Press, Baltimore.

88. Lee S, Aronoff S (1967). Boron in Plants: A Biochemical role. Science 158(3802): 798-799. Lexander OA, Baumann CA (1966). Selenium Metabolism VI. Effect of Arsenic on the Excretion of Selenium in the Bile. Toxicol. Aplied Pharmacol. 9: 106-115.

89. Levander O (1983). Considerations in the design of selenium bioavailability studies. Fed. Proc. 42: 1721-1725.

90. Lewin J, Reimann BEF (1969). Silicon and plant growth, Ann. Rev. Plant Physiol. 20: 289-304.

91. Lewis OH (1980). Boron is an essential micronutrient for higher plants. Plant Physiol. 113: 649-555.

92. Malhotra VK (1998). Biochemistry for Students. Tenth Edition. Jaypee Brothers Medical Publishers (P) Ltd, New Delhi, India.

93. McDowell LR, Conrad JH, Ellis GI (1984). Mineral deficiencies and imbalances and their diagnosis. Proc. Symp. on Herbivore nutrition in subtropics and tropics, Pretoria pp. 67-88.

94. Meeker JD, Rossano MG, Protas B, Diamond MP, Puscheck E, Daly D, Paneth N, Wirth JJ (2008). Cadmium, lead and other metals in relation to semen quality: Human evidence for Molybdenum as a male reproductive toxicant. Environ. Health Perspec. 116: 1473-1479.

95. Merck VM (1986). The Merck Veterinary Manual. Sixth Edition. A handbook of diagnosis, therapy and disease prevention and

control for the veterinarian. Published by Merck and Co., Inc., Rahway, New Jersey, USA.

96. Mertz W (1974). Chromium as a dietary essential for man. In: WG Hoekstra *et al* eds., Trace Element Metabolism in Animals, 2nd ed. University Park Press, Baltimore.

97. Messer HH, Armstrong WD, Singer L (1974). Essentiality and function of fluoride. In WG Hoekstra et al., eds., Trace Element Metabolism in Animals, 2nd ed. University Park Press, Baltimore.

98. Migon C, Morelli J, Nicolas E, Copin-Motegut G (1991). Evaluation of total atmospheric deposition of Pb, Cd, Cu and Zn to the Ligurian sea. Sci. Total Environ. 105: 135-148.

99. Mongia AD (1966). Phosphorus-zinc relationship in wheat as affected by sources and rates of applied phosphorus M.Sc. Thesis. Punjab Agricultural University, Hissar.

100. Murray RK, Granner DK, Mayes PA, Rodwell VW (2000). Harper's Biochemistry, 25th Edition, McGraw-Hill, Health Profession Division, USA.

101. Nandi D, Patra RC, Swarap D (2006). Oxidative stress indices and plasma biochemical parameters during oral exposure to arsenic in rats. Food Chem. Toxicol. 44: 1579-1584.

102. Neher GM, Doyle LP, Thrasher DM, Plumlee MP (1956). Radiographic and histopathological findings in the bones of swine deficient in manganes. Am. J. Vet. Res. 17: 121-128.

103. Nelson TS, Ferrara LW, Stover HL (1968). Effect of phytate on the calcium requirement of chicks. Poult. Sci. 47: 185-189.

104. Nielsen FH (1976). In: Trace Elements in Human Health and Disease. A Prasad, Editor, Academic Press, New York 2: 379-399.

105. Nielsen FH (1974). Essentiality and function of nickel. In: WG Hoekstra *et al* eds., Trace Element Metabolism in Animals, 2nd ed. University Park Press, Baltimore.

106. Nielsen FH, Ollerich DA (1973). Studies on a vanadium deficiency in chicks, Fed. Proc. 32: 929.

107. Nwadiaro PO, Nwachukwu I (2007). Inhibition of Pathogenic Fungi by Ethnobotanical Extracts of *Cymbopogon citrates, Ceiba pentandra* and *Loranthus bengwelensis*. Niger. J. Biotechnol. 18: 61-65.

108. Nwokolo EN, Bragg DB (1977). Influence of phytic acid and crude fibre on the availability of minerals from four protein supplements in growing chicks. Can. J. Anim. Sci. 57: 475-477.

109. O' Connor BB (1995). Vernacular health care response to HIV and AIDS. Alternative Therapies 1(5): 35-52.

110. Ogbonna AI, Makut MD, Gyar SD, Adamu EU (2007). Antimicrobial Activity of ethanolic extracts of the seeds of Ricinus communis L. Niger. J. Biotechnol. 18(1-2): 40-43.

111. Olaiya CO (2006). Effects of three plant bioregulators on some biochemical properties of *Lycopersicon esculentum* (L.) Mill. Ph.D Thesis, Department of Biochemistry, University of Ibadan, Nigeria.

112. Osagie AU (1998). Nutritional quality of plant foods pp. 221-244.

113. Ozcan M (2003). Mineral Contents of some Plants used as condiments in Turkey. Food Chemistry 84: 437-440.

114. Pal R, Nath R, Gill DK (1993). Influence of Ethanol on cadmium accumulation and its impact on lipid peroxidation and membrane- Soetan et al. 221 bound functional enzymes (Na+/ K+ ATPase, acetylcholinesterase) in various regions of adult rat brain. Neurochem. Int. 235: 451-458.

115. Panzek J, Ostadalova I (1967). The Protective effect of small amounts of selenite in sublimate intoxication. Experientia, 23: 142-143.

116. Pathak P, Kapil U (2004). Role of trace elements zinc, copper and magnesium during pregnancy and its outcome. Indian J. Paediatr. 71: 1003-1005.

117. Patwardhan VN (1961). Nutrition in India. Ind. J. Med. Sci., Bombay. Pierson RE, Aenes WA (1958). Treatment of chronic copper poisoning in sheep. J. Am. Vet. Med. Ass. 133: 307-311.

118. Pirkle JL, Kaufmann RB, Brody DJ, Hickman T, Gunter EW, Paschal DC (1998). Exposure of the US population to lead, 1991-1994. Environ. Health Perspec. 106: 745-750.

119. Pollack M, Leeuwenburgh C (1999). Mollecular mechanisms of oxidative stress in aging: free radicals, aging, antioxidant and disease. In: Handbook of Oxidants and Antioxidants in Exercise

(Sen CK, Packer L, Hinninen O. Eds). Elsevier Science, BV, pp. 881-923.

120. Rajaratnam JA, Lowry JB (1974). The role of boron in the oil palm (*Elaeis guineensis*), Ann. Bot. 38: 193-2000.
121. Ramasarma T, Crane FL (1981). Does vanadium play a role in cellular regulation? Current Topics in Cellular Regulation 20: 247-301.
122. Rao CN, Rao BSN (1981). Trace element content of Indian foods and the dietaries. Indian J. Med. Res. 73: 904-909.
123. Roberts JV (1985). Textbook of Biology, First Edition, JB Publishers, India.
124. Roesijiadi G, Robinson WE (1994). Metal regulation in aquatic animals: mechanism of uptake, accumulation and release. In: Malins DC, Ostrander GK (Eds), Aquatic Toxicology (Molecular, Biochemical and Cellular Perspectives), Lewis Publishers,London p. 539.
125. Roginski EE, Mertz W (1969). Effects of chromium supplementation on glucose and amino acid metabolism in rats fed a low protein diet, J. Nutr. 97: 525-530.
126. Romeo M, Gnassia-Barelli M (1988). *Donax trunculus* and *Venus verrucusa* as bioindicators of trace metal concentrations in Mauritanian coastal waters. Mar. Biol. 99: 223-227.
127. Romero R (2003). Intra-uterine infection, premature birth and the Fetal Inflammatory Response Syndrome. J. Nutr. 33: 1668-1673.
128. Rush D (2000). Nutrition and Maternal mortality in the developing world, Am. J. Clin. Nutr. 72(Suppl.): 2125-2405.
129. SAtoskar RS, Bhandarkar SD, Alnapure SS (1999). Pharmacology and Pharmacotherapeutics, 6th Edition, Popular Prakashan, Mumbai, India pp. 1012-1013.
130. Sadrzadeh SM, Saffari Y (2004). Iron and brain disorders. Am. J. Clin. Pathol, pp. 64-70.
131. Sandel EB (1959). Colorimetric Determination of Traces of Metals. 3rd Edition, Interscience Publishers Inc., New York p. 460.
132. Sardesai OK (1993). Molybdenum: An essential trace element. Nutr. Clin. Pract. 8: 277-281.
133. Sarker S, Yadar P, Trivedi R, Bansal AK, Bhatnagar D (1995). Cadmium-induced lipid peroxidation and the status of the

antioxidant system in rat tissues. J. Trace Elem. Med. Biol. 9(3): 144-149.

134. Schrauzer GN (2000). Selenometionine: A role of its nutritional significance, metabolism and toxicity. J. Nutr. 130: 1653-1656.
135. Schwarz K (1974). New essential trace elements (Sn, V, F, Si): Progress report and outlook. In: WG Hoekstra *et al* eds., Trace Element Metabolism in Animals, 2nd ed. University Park Press, Baltimore.
136. Semadi A, Deruelle S (1993). Lead pollution monitoring by transplanted lichens in Annaba area (Algeria). Rev. Poll. Atmos. Oct.-Dec. pp. 86- 102.
137. Sharma S, Mahotra P, Bhattacharyya AK (2008). Effect of electroplating industrial waste on "available phosphorus" of soil in relation to other physico-chemical properties. Afr. J. Environ. Sci. Technol. 2(9): 257- 264.
138. Sherman RA (1990). Influence of iron on immunity and disease resistance. Ann. N.Y. Acad. Sci. 587: 123-139.
139. Shkol'nik MY, Abysheva LN (1975). Effect of boron deficiency on the level of the growth inhibitor flavonol-3-glycoside and other flavonoids in tomatoes, Fiziol. Biokhim. Kul't. Rast. 7: 291-297.
140. Sidhu P, Gorg ML, Morgenstern P, Vogt J, Butz T, Dhawan DK (2004). Role of Zinc in regulating the levels of hepatic elements following nickel toxicity in rats. Biological Trace Elements Res. 102: 161-172.
141. Simsek A, Aykut O (2007). Evaluation of the microelement profile of Turkish hazelnut (*Corylus avellana* L) varieties for human nutrition and health. Int. J. Food Sci. Nutr. 58: 677-688.
142. Smith DM Jr, Mielke HW, Heneghan JB (2008). Subchronic lead feeding study in male rats. Arch. Environ. Contamn. Toxicol. 55(3): 518-528.
143. Soetan KO (2008). Pharmacological and other Beneficial effects of Antinutritional Factors in Plants- A Review. Afr. J. Biotechnol Vol. 7(25): 4713-4721.
144. Soetan KO, Oyewole OE (2009). The need for adequate processing to reduce the anti- nutritional factors in plants used for human food and animal feeds- A Review. Afr. J. Food Sci. 3(9): 123-132.

145. Sokol RZ, Saixi W, Yu-Jui Y, Frank ZS, Elizabeth G, Robert EC (2002). Long-term, low-dose lead exposure alters the gonadotropin-releasing hormone system in the male rat. Environmental Health Perspectives 110: 871-874.

146. Southern LL, Baker DH (1983). *Eimeria acervulina* infection in chicks fed deficient or excess levels of manganese, J. Nutr. 113: 172-177.

147. Spallholz EJ, Boylan ML, Larsen SH (1990). Advances in understanding selenium's role in the immune system. Ann. N.Y. Acad. Sci. 587: 123- 139.

148. Strasia CA (1971). Vanadium: Essentiality and toxicity in the laboratory rat. Ph.D Thesis, University Microfilms, Ann. Arbor, Mich.

149. Streeten DHP, Williams EMV (1952). Loss of cellular potassium as a cause of intestinal paralysis in dogs. J. Physiol. 118: 149-170.

150. Surai PF (2000). Selenium in poultry nutrition 2. Reproduction, egg and meat quality and practical applications. World Poult. Sci. J. 58: 431- 450.

151. Szabo G, Chavan S, Mandrekar P, Catalano D (1999). Acute alcoholic consumption attenuates IL-8 and MCP-1 induction in response to *ex vivo* stimulation. J. Clin. Immunol. 19: 67-76.

152. Tan JC, Burns DL, Jones HR (2006). Severe ataxia, myelopathy and peripheral neuropathy due to acquired copper deficiency in a patient with history of gastrectomy. J. Paenteral Nutr. 30: 446-450.

153. Taneja V, Mishra K, Agarwal KN (1986). Effect of early iron deficiency in the rat on gamma amino butyric acid shunt in brain. J. Neurochem. 46: 1670-1674.

154. Tejada-Jimenez M., Galvan A., Fernandez E., Liamas A. (2009). Homeostasis of the micronutrients Ni, Mo and Cl with specific biochemical functions. Curr. Opin. Plt. Biol. 12: 358-363.

155. Telisman S, Colak B, Pizent A, Jurasovic J, Cvitkovic P (2007). Reproductive toxicity of low-level lead exposure in men. Environ. Res. 105: 256-266.

156. Thilly CH, Vanderpas JB, Bebe N, Ntambue K, Contempre B, Swennen B, Moreno-Ryes R, Bourdoux P, Delange F (1992). Iodine deficiency, other trace elements and goitrogenic factors

in the ethiopathogeny of iodine deficiency disorders (IDD). Biological Trace Elem. Res. 32(1- 3): 229-243.

157. Underwood EJ (1971). Trace Elements in Human and Animal Nutrition, 3rd Edition, Academic Press, New York p. 116.

158. Uppala RT, Roy SK, Tousson A, Barnes S, Uppala GR, Eastwood Da (2005). Induction of cell proliferation micronuclei and hyperdiploidy /diploidy in the mammary cells of DDT and DMBA-treated pubertal rats.Environmental and Molecular Mutagenesis 46: 43-52.

159. Usero J, Morillo J, Gracia I (2005). Heavy metal concentrations in molluscs from the Atlantic coast of Southern Spain. Chemosphere 59: 1175-1181.

160. Vobecky M, Babicky A, Lener J, Svandova E (1996). Interaction of bromine with iodine in the rat thyroid gland at enhanced bromide intake. Biological Trace Element Res. 54(3): 207-212.

161. Wang L, Xun P, Zhao Y, Wang X, Qian L, Chen F (2008). Effects of lead exposure on sperm concentrations and testes weight in male rats: a meta-regression analysis. Toxicol. Environ. Health. 71(7): 454- 463.

162. Williams US (1992). A Textbook of Biology, Third Edition. Wennberg A (1994). Neurotoxic effects of selected metals. Scand. J. Work. Environ. Health 20: 65-71.

163. Wilcox LV (1960). Boron injury to plants. Agric. Info. Bull. 211, US. Dept of Agric., Washington, DC, 1960.

164. Wood RJ (2000). Assessment of marginal zinc status in humans. J. Nutr. 130: 1350-1354.

165. World Health Organization (WHO) (1992). Environmental Health Criteria, Cadmium, WHO, Geneva p.134.

166. Wycherly BJ, Moak MA, Christensen MJ (2004). High dietary intake of sodium selenite induces oxidative DNA damage in rat liver. Nutr. Cancer 48: 78-83.

167. Youdim MBH, Ben Shacher D, Yehuda S (1989). Putative biological mechanism of the effect of iron deficiency on brain biochemistry and behavior. Am. J. Clin. Nutr. 50: 607-650.

# Citations

## CHAPTER 1

Akl C. Fahed, Abdul-Karim M. El-Hage-Sleiman, Theresa I. Farhat, and Georges M. Nemer, "Diet, Genetics, and Disease: A Focus on the Middle East and North Africa Region," Journal of Nutrition and Metabolism, vol. 2012, Article ID 109037, 19 pages, 2012, doi:10.1155/2012/109037.

## CHAPTER 2

Chantal Bémeur, Paul Desjardins, and Roger F. Butterworth, "Role of Nutrition in the Management of Hepatic Encephalopathy in End-

Stage Liver Failure," Journal of Nutrition and Metabolism, vol. 2010, Article ID 489823, 12 pages, 2010. doi:10.1155/2010/489823.

# CHAPTER 3

J. Dias and R. Ortiz, "Transgenic Vegetable Breeding for Nutritional Quality and Health Benefits," *Food and Nutrition Sciences*, Vol. 3 No. 9, 2012, pp. 1209-1219. doi: 10.4236/fns.2012.39159.

# CHAPTER 4

Ana Rosa Cunha, Bianca Umbelino, Margarida L. Correia, and Mario Fritsch Neves, "Magnesium and Vascular Changes in Hypertension," International Journal of Hypertension, vol. 2012, Article ID 754250, 7 pages, 2012, doi:10.1155/2012/754250.

# CHAPTER 5

Pooja Parmar, Nilima Kumari, and Vinay Sharma, Structural and Functional Alterations in Photosynthetic Apparatus of Plants under Cadmium Stress, doi: 10.1186/1999-3110-54-45.

# CHAPTER 6

Stig Bengmark, Integrative Medicine and Human Health - the Role of Pre-, Pro- and Synbiotics, doi:10.1186/2001-1326-1-6.

# CHAPTER 7

T. S. Workneh and G. Osthoff, A Review on Integrated Agro-technology of Vegetables, ISSN 1684–5315 © 2010.

# CHAPTER 8

K. O. Soetan, C. O. Olaiya, and O. E. Oyewole, The importance of mineral elements for humans, domestic animals and plants: A review, ISSN 1996-0794.

# Index


**A**

Adenosine triphosphate (ATP) 114, 239, 256, 257
Adrenocorticotropic hormone (ACTH) 258
Aminoethoxyvinylglycine (AVG 196
Aminolevulinate (ALA) 133
Anabolic reaction 132
Angiotensin Converting Enzyme (ACE) 19

**B**

Beta-lactoglobulin (BLG) 164
Biochemical change 183, 196, 200, 201, 202, 207, 212, 213, 226
Body mass index (BMI) 56
Brain-Derived Neurotrophic Factor (BDNF) 18

**C**

Cardiovascular diseases 111, 112, 117, 122
Cement production 130
Central nervous system (CNS) 268
Colorectal cancer (CRC) 26
Colorectal distension (CRD) 164
Controlled-atmosphere packaging (CAP 210

Controlled- atmosphere storage (CAS) 210
Coronary Artery Disease (CAD) 20

**D**

Days after treatment (DAT) 132
Digalactosyldiacyl-glycerol (DGDG) 132

**E**

Ethylene production 194, 195, 196, 197, 202, 211, 215, 216, 217

**F**

Fructo-oligosaccharides (FOS) 157
Fructooligosacharides (FOS) 156

**G**

Gastrointestinal tract (GIT) 263
Genetically modified (GM) 85
"Glucose tolerance factor" (GTF) 261
Glutathione S-transferases (GSTs 27
Glycolytic enzyme glyceraldehyde-3-phosphate dehydrogenase (GAPDH) 60

**H**

Heat shock proteins (HSP-72) 159
Hepatic encephalopathy (HE) 50
Hepatitis E virus (HEV) 96
Hexaphosphoric acid 257
High density lipoprotein (HDL) 274
High density polyethylene (HDPE 218
Hyperhomocysteinemia (HHC) 25

**I**

Indocyanine green (ICGR15) 168
Iodine deficiency disorders (IDD) 276, 298
Irritable bowel syndrome (IBS) 158, 159, 176

**L**

Lipooxygenase (LOX) 132
Lipoprotein Lipase (LPL) 20
Low density polyethylene (LDPE) 218
Lycopene epsilon cyclase (LCY-e) 86

**M**

Micronutrient deficiencies (MNDs 3, 18
Mid-arm circumference (MAC) 57
Mid-arm muscle circumference (MAMC) 57
Middle East and North Africa (MENA) 1, 3
Model for end-stage liver disease (MELD) 56
Modified atmosphere packaging (MAP 210, 211, 212, 213, 214
Modified Atmosphere Packaging (MAP 210
Monogalactosyldiacyl-glycerol (MGDG) 132

## N

N-acetylcysteine (NAC) 66
Neomycin phosphotransferase II (NPTII) 91
Neural tube defects (NTDs 27
Nicotinamide-adenine-dinucleotide- (NAD 28
Nitric oxide (NO) 113
NRAMP (natural resistance associated macrophage protein) 131

## O

Oxidation-reduction potential (ORP 207

## P

Parenteral nutrition (PN) 166
Peroxisome Proliferator-Activated Receptor Gamma (PPARG) 20
Phosphatidyl glycerol (PG) 132
Polygalacturonase (PG 202
Polystyrene (PS 218
Polyvinylchloride (PVC 218
Polyvinylidine chloride (PVDC 218
Potentially toxic 129, 130
Pulse wave velocity (PWV) 116
Pylorus-preserving pancreatoduodenectomy (PPPD) 167

## Q

Quantitative trait loci (QTL) 87

## R

Red blood cells (rbc) 263

## S

Selenoenzymes glutathione peroxidase (GSH-Px) 269
Sequence characterized amplified repeat (SCAR) 95
Severe acute pancreatitis (SAP) 165
Sex-linked anemia (Sla 25
Short chain fatty acids (SCFAs) 153
Spontaneously hypertensive rats (SHRs) 115
Subjective Global Assessment (SGA) 57
Superoxide dismutase (SOD) 269

## T

Tetrahydrofolate (THF) 25
Tomato fruit 185, 189, 196, 199, 222, 231, 232
Triceps skin fold thickness (TST) 57
Tumor necrosis factor-alpha (TNF- ) 55
Tumor necrosis factor (TNF) 162

## U

Ultraviolet B (UVB) 22

## V

Vinyl acetate (VA 218
Vitamin D Deficiency (VDD) 10, 22